Industrial Applications
of Microbiology

Jacques Rivière

Professor, l'Institut National Agronomique, France

Translated and edited with additional
material by

M. O. Moss

Senior Lecturer, Department of Microbiology, University of Surrey

and

J. E. Smith

Professor, Department of Microbiology, University of Surrey

A HALSTED PRESS BOOK

John Wiley & Sons
New York

Published in the U.S.A. and Latin America
by Halsted Press, a Division
of John Wiley & Sons, Inc.

This edition first published by Surrey University Press, 450 Edgware Road, London W2 1EG

Originally published in French under the title *Les Applications Industrielles de la Microbiologie*

© Masson et Cie, Paris 1975

English translation and chapters 6 and 7 copyright © M. O. Moss and J. E. Smith 1977

This edition first published 1977

Library of Congress Cataloging in Publication Data
Rivière, Jacques.
Industrial applications of microbiology.
Translation of Les applications industrielles de la
microbiologie.
"A Halsted Press book."
Bibliography: p.
Includes index.
1. Industrial microbiology. I. Title.
QR53.R5413   660'.62   77-22815
ISBN 0-470-99265-4

Printed in Great Britain by Thomson Litho Ltd, East Kilbride, Great Britain

Contents

Introduction

The title of this book has been retained from the work by Professor Jacques Rivière on which it is based, but it is necessary first to point out that the full range of industrial applications of microbiology is too great to be encompassed in a text of this size. It would include subjects as diverse as the production of vaccines and hyperimmune sera, food preservation methods, the monitoring and control of microbial deterioration in such other spoilable products as textiles, timber, paints, fuels, cosmetics and pharmaceuticals, the production of potable water from impure sources, the treatment of microbiologically undesirable industrial and agricultural wastes and, in industrialized agriculture, the control of infective animal and crop diseases and the microbiological conservation of soil fertility. It would also require consideration of a great range of processes which, though involving complex microbial activities, are essentially scaled up versions of former domestic arts, such as the industrial production of cheeses, yoghurt, vinegar, fermented pickles, sauerkraut, soya sauce and other Oriental condiments, alcoholic beverages, the retting of plant fibres, leather tanning and the post-harvest maturation of tea, coffee, olives and tobacco.

We are concerned here with industrial microbiology in a more restricted sense, namely the study of certain processes whereby microorganisms grown on a large scale, almost invariably in liquid media, are used to effect conversion of low cost substrates to commercially valuable substances or materials, which may then be extracted. The industrial microbiologist loosely terms all such processes 'fermentations' and that branch of industry which employs them is called the fermentation industry.

The scientific fermentation industries made their appearance early in the present century, but had their earlier origins in traditional fermentation processes such as brewing and wine making which had been carried out for many hundreds of years. In some cases, these processes had been developed to an industrial scale without any inkling of their essentially biological nature. In Britain, for example, brewing became partly industrialized as early as the end of the fifteenth century, when the newly introduced practice of hopping gave, for the first time, a product which had keeping qualities sufficient for it to be produced in bulk and distributed to the alehouses, instead of having to be brewed on the premises: charters were granted to Companies of 'Common Brewers' and in 1767 Michael Cambrune of the Worshipful Company of Brewers was describing fermentation as

> 'a sensible internal motion of the particles of a mixture: by the continuance of this motion, the particles are gradually removed from their former situation and, after some visual separation, are joined together in different order and arrangement, so as to constitute a new compound.'

By this time, Van Leeuwenhoek (1632–1723) had already observed yeasts and other organisms microscopically, but it was not until the early nineteenth century that Cagniard-Latour and Schwann noticed that the fermentation of wine and beer was accompanied by the development of yeast cells, called 'Zuckerpilz' by Schwann and *Saccharomyces* by another worker, Meyen. It was Pasteur who was to demonstrate, later in the century, the relationship between the microbial agents which caused fermentations and those which were responsible for human and animal diseases. The medical value of the work of Pasteur and his successors at the Pasteur Institute is sometimes allowed to overshadow the importance of Pasteur's conclusive demonstration of the microbiological nature of fermentation, and of the many studies he made of the fermentation processes of his time, showing how accidental contaminations or infections of wines and beer might be controlled and efficiency of fermentation improved. Seventy years later, also at the Pasteur Institute, Monod was to derive, from the results of small scale laboratory experiments, equations describing the growth of bacteria in liquid cultures which are of fundamental applicability in industrial fermentations.

The idea of adapting the kind of microbial fermentations studied by Pasteur to produce specific chemical substances for use as raw materials in other manufacturing processes, or for direct sale, was slow to develop. In the 1880's, lactic acid was being produced by fermentation as a substitute for tartaric acid in baking powders, but the large scale production of raw chemicals by fermentation did not begin until the First World War. Then, Germany produced glycerol by fermentation for use in the manufacture of explosives; production of acetone, also for explosives manufacture, by the acetone-butanol fermentation was developed in Britain and later moved to Canada and the United States where grain starch, the original substrate, was replaced by molasses; industrial plants for producing ethanol, as a fuel, by fermentation and distillation were constructed in a number of countries.

Between the wars, development was slower but *n*-butanol, an unwanted by-product of the wartime acetone fermentation, together with its derivative, *n*-butyl acetate, found an important use as a lacquer solvent in the growing motor vehicle industry. Methods for producing citric and gluconic acids by fermentation were devised.

In the Second World War, production of ethanol, acetone and other volatile chemicals by fermentation declined because of the increasing cost of shipping molasses, together with losses by enemy action. Chemical means of producing these substances from petroleum were becoming more effective. The immediate post-war years saw the final demise of the large industrial solvent fermentation-distillation plant. However, during the war a new phase had begun, with the manufacture by fermentation of a vitamin, riboflavine, clinical dextrans and the first antibiotic, penicillin, to be followed after the war with the manufacture by fermentation of a host of newly developed antibiotics, together with many enzymes, amino acids, vitamins including B_{12}, steroids and other pharmacologically active compounds. A new subject, microbial genetics, began to receive energetic study in the hope that it would reveal ways of rapidly developing high-yielding industrial strains.

It is significant that, in the post-war decades, rising populations, food shortages and growing international distress at the inequalities of life-standards between nations should have shifted the emphasis towards the use of fermentation methods for producing economical protein foodstuffs from low-cost raw materials. These methods are currently being used mainly to produce animal feed supplements but are likely to be extended to the manufacture of food substances for human consumption.

Professor Rivière developed the book on which the present volume is based from material contained in a lecture course on industrial microbiology which, over a number of years, he had organized at the French Institut National Agronomique. In the present version, care has been taken to preserve the essential features of the original text: a mix of immediate practicalities with those fundamentals which have to be carefully explained to the student, a recognition that industrial microbiology is a rapidly advancing subject with few fixed points, and an attempt to summarize and profit from the lessons of past and present experiences. Freedom has been exercised in translation; some portions have been omitted, others have been altered or expanded. Most of Professor Rivière's Tables have been retained, in some cases slightly modified. They are a characteristic and essential feature of the book, derived from contemporary publications and from Professor Rivière's own work: source references are generally not given since the tabulations are designed merely to exemplify the present state of the art.

Chapter 1 explains the meaning of fermentation in relation to industrial processes, provides the essential biochemical background and describes how strains of microorganisms having potential usefulness are isolated, maintained and improved. Chapter 2 deals with the formulation of culture media, including the energy sources and the important macro- and micro-nutrient substances, and with the sterilization of these media or their constituents by heat.

Control of such fermentation conditions as temperature, pH and aeration, and the different methods of growing microorganisms in batch, continuous and dialysis culture, are examined in Chapter 3, together with various approaches to extraction of the desired product from the final culture.

In the remaining chapters, various groups of products manufactured by microbiological means are discussed—microbial proteins, such diverse products of microbial metabolism as polyols, industrial solvents, organic acids, vitamins, nucleotides, polysaccharides and enzymes. The original text has been augmented by chapters on antibiotics and fermented alcoholic beverages.

This edition of a work hitherto unavailable in English is intended primarily for the student, undergraduate and postgraduate, as a simple treatise combining both practical and fundamental aspects of industrial microbiology. Nonetheless, it is hoped that the more experienced worker seeking broad information will find, at least, some food for thought in these pages.

M. O. Moss
and
J. E. Smith
1977

1 Fermentation: Biochemistry and Culture Improvement

1.1 Definition of industrial fermentation

The commercial exploitation of microorganisms encompasses the fermentation and allied industries. The word fermentation comes from the Latin verb *fervere* which means to boil, and is derived from the fact that the start of alcoholic fermentation is characterized by bubbles of gas continuously bursting at the surface, giving the must or wort the appearance of a boiling liquid. The definition has been extended to include the numerous oxidation-reduction reactions in which organic compounds, used as sources of carbon and energy, act as acceptors or donors of hydrogen ions. Substrates which accept hydrogen ions give rise to the products of fermentation which accumulate in the medium. They are numerous and many of them result from the metabolism of sugars. The large variety of possible terminal products is shown, for example, by the diversity of pathways by which pyruvic acid can be reduced (Figure 1.1).

Which of these compounds is actually produced depends on the presence or absence of particular enzymes in the microorganism. Thus the lactic acid bacteria do not produce pyruvate decarboxylase, but reduce pyruvic acid to lactic acid, whereas beer yeasts can decarboxylate pyruvic acid reducing the resulting C_2 compound to ethyl alcohol.

The initial or primary pathways by which compounds such as glucose are metabolized are common to the cells of many microorganisms and the better understood of these will now be discussed.

The anaerobic metabolism of glucose may occur via the Embden-Meyerhof-Parnas pathway (Figure 1.2).

A second process for the catabolism of glucose is the Entner-Doudoroff pathway found in many bacteria of the genus *Pseudomonas* (Figure 1.3).

The pentose cycle, which may form a loop incorporated into the EMP, is found, for example, in the heterolactic fermentation of *Leuconostoc mesenteroides* (Figure 1.4).

2

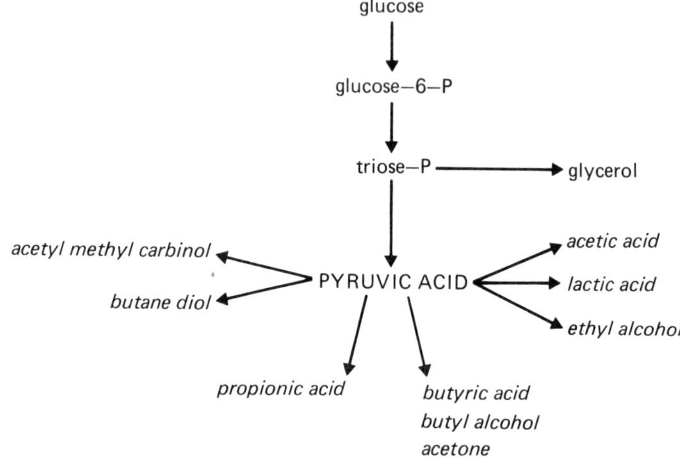

Figure 1.1   Fermentation products from the reduction of pyruvic acid

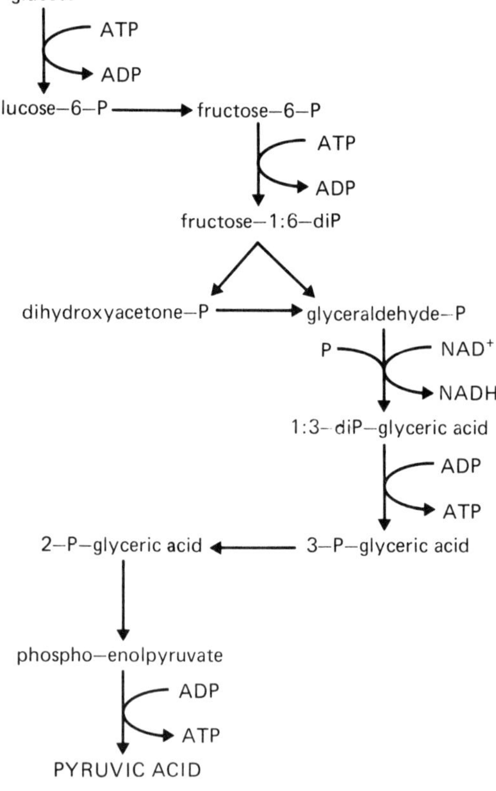

Figure 1.2   Catabolism of glucose via the Embden-Meyerhof-Parnas pathway
(EMP or glycolysis)

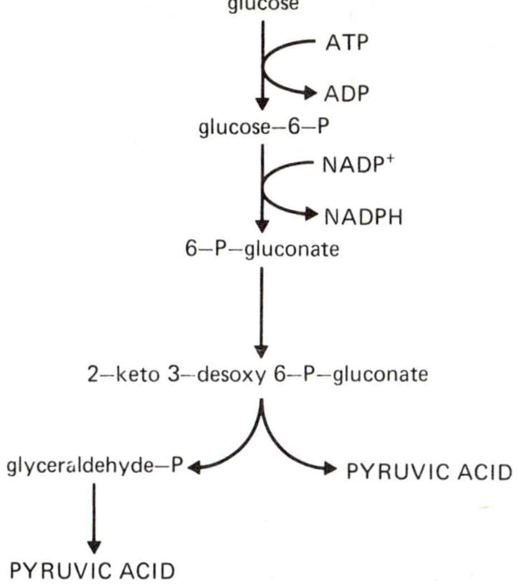

Figure 1.3   The Entner-Doudoroff pathway

Any particular microorganism may use one or more of these several catabolic pathways. Thus osmophilic yeasts, such as *Saccharomyces rouxii*, produce from glucose a mixture of alcohols including ethanol, glycerol, arabitol and erythritol all of which arise from intermediates of the EMP pathway and the pentose cycle (Figure 1.5).

The study of fermentation products is useful for both the identification and classification of bacteria. Thus strains of the genus *Lactobacillus* may be classified into two groups:

Homofermentative lactobacilli which are used for the production of lactic acid by the fermentation of glucose. Some strains produce nearly 100% lactic acid from glucose via the EMP pathway. They possess an active aldolase, which splits fructose diphosphate into two molecules of a triose phosphate.

Heterofermentative lactobacilli which produce ethanol, glycerol, acetic acid and carbon dioxide as well as lactic acid. Because they do not synthesize an aldolase, degradation of glucose proceeds via the pentose pathway as far as the formation of ribulose-5-phosphate (Figure 1.6).

The reactions described so far are anaerobic and are properly described as fermentations but industry generally uses the word fermentation to embrace all processes in which modifications of a chemical structure are brought about by enzymes of microbial origin. Thus, even aerobic processes, such as the production of gluconic acid from glucose by cultures of the mould *Aspergillus niger* or of dihydroxacetone from glycerol by various species of the bacterial genus *Acetobacter* (*A. xylinum* or *A. suboxydans*), are often loosely described as fermentations.

1.2   Search for new cultures

The search for suitable cultures can thus be viewed as an attempt to discover species which possess the array of enzymes and other properties for producing the desired product in high yield.

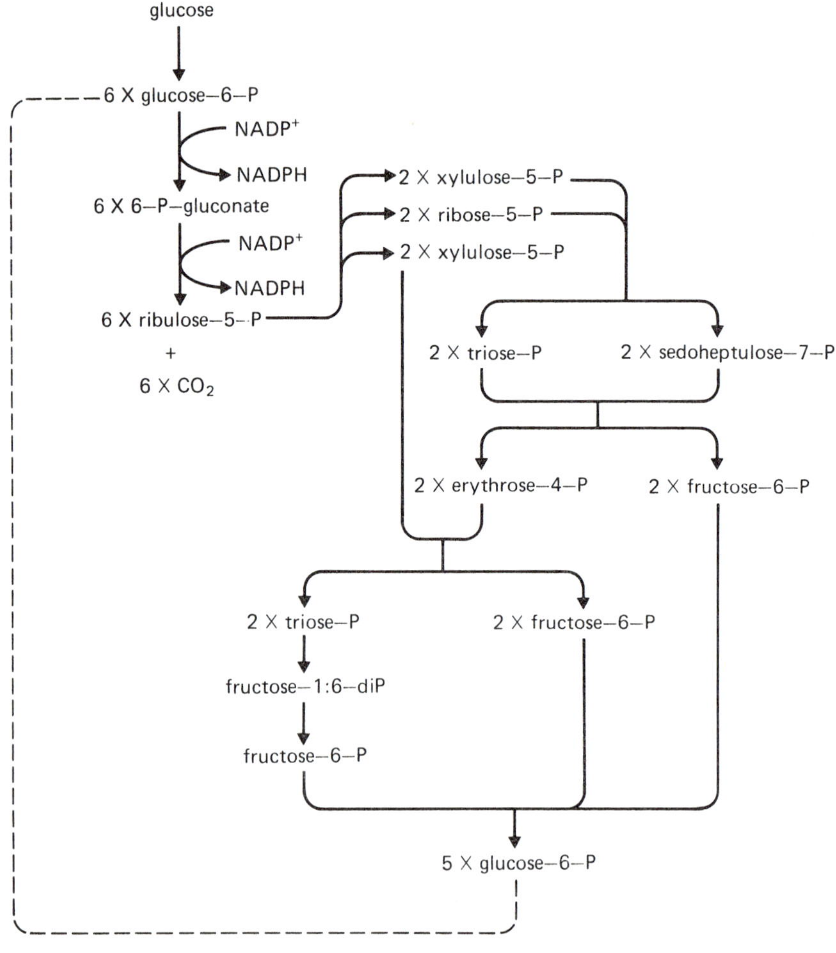

Figure 1.4   The pentose cycle

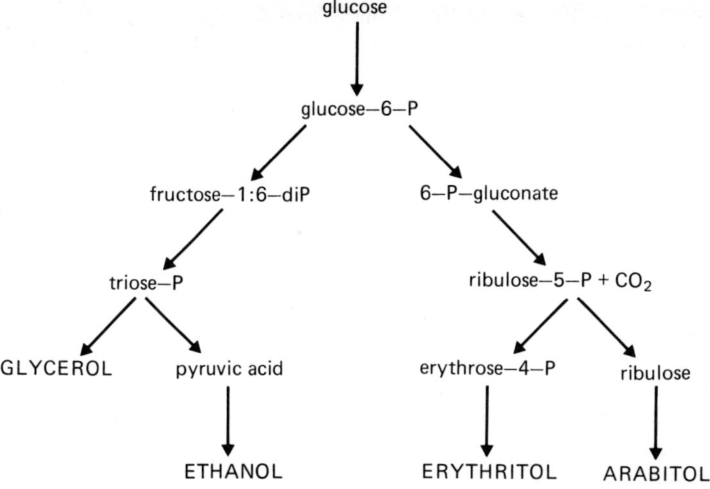

Figure 1.5   Production of alcohols by osmophilic yeasts

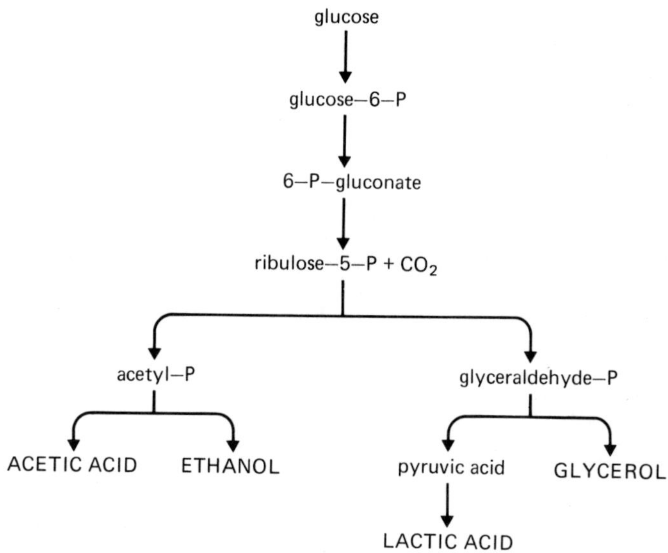

Figure 1.6   Heterolactic fermentation

The discovery of new cultures may result from chance observations as was the case of Fleming's culture of *Penicillium notatum*, isolated from an accidental contamination of an agar plate seeded with a culture of *Staphylococcus aureus*. But there are general methods for the systematic search for new cultures.

Established culture collections may furnish lists of strains of known characteristics. If, for example, a cellulase is required, a range of cellulolytic strains of bacteria and fungi may be obtained (Table 1.1). A more complete list of culture collections throughout the world is provided by Hesseltine, C. W. and Haynes, W. C. (1973).

It may be useful to isolate directly from a natural source (air, water or soil) or from decomposing organic material. The culture of *Penicillium chrysogenum*, used for the industrial production of penicillin, was isolated from a mouldy melon bought in Peoria, and several cultures of *Bacillus* (*B. subtilis* and *B. licheniformis*), used for the production of alkaline proteases, were isolated from hippopotamus dung from Copenhagen zoo.

Table 1.1   Major culture collections

Name	*Address*	*Type of microorganism*
American Type Culture Collection (ATCC)	Rockville, Maryland 20852, USA	bacteria, fungi, actinomycetes, algae, protozoa
Central Bureau voor Schimmelculture, (CBS)	Baarn, Holland	fungi, actinomycetes
Commonwealth Mycological Institute, (CMI)	Kew, England	fungi
Agricultural Research Service Culture Collection (NRRL)	Peoria, Illinois 61604 USA	bacteria, fungi, actinomycetes
National Collection of Industrial Bacteria (NCIB)	Aberdeen, Scotland	bacteria
Centre de Collection de Types Microbiens	Lausanne, Switzerland	bacteria
Culture Collection	Cambridge University, England	algae and protozoa
National Collection of Type Cultures (NCTC)	London, England	bacteria, fungi, actinomycetes
Culture Collection of Indiana University	Bloomington, Indiana 47405 USA	algae

Rapid screening techniques for testing the properties of a heterogeneous microbial population, such as that occurring in a sample of soil, should preferably combine simultaneous isolation and selection.

The crowded plate technique

This simple method, used in isolating antibiotic producers, consists of spreading a series of dilutions of a soil suspension on the surface of nutrient agar plates. Those plates which give 100–300 colonies after incubation are flooded with a suspension of the pathogenic culture against which an antibiotic is required. After further incubation, those colonies from the soil suspension which produce characteristic zones of inhibition in the confluent growth of the test organism are isolated, purified and tested for their ability to produce activity in liquid medium against the pathogen.

Auxanography

This technique is mainly used for detecting cultures producing growth factors. A strip of filter paper (1·5 × 12 cm) is placed across the bottom of a Petri dish in such a way that the two ends pass over the edge of the dish. A disc of the same paper is placed over it covering the bottom of the dish and nutrient agar is poured on. When it has set, a sample from a soil dilution is spread on the surface in such a way as to give colonies sufficiently separated from each other as to be easily identified. A separate pour plate is prepared of nutrient agar inoculated with the indicator organism. This is usually an auxotroph requiring a specific growth factor, amino acid or vitamin (in which case it is inoculated into a minimal medium lacking this factor); alternatively it may be a culture for which an antibiotic is required (which is inoculated into a complete medium). When the indicator plate has set, the first plate of agar is carefully and aseptically lifted out with the help of tweezers and a spatula and placed, without inverting it, onto the surface of the second.

Metabolites diffuse from the first layer of agar into the layer containing the indicator organism and colonies to be sub-cultured may be recognized by the production of a zone of growth of the auxotroph (in the production of a growth factor) or by a zone of inhibition (in the case of an antibiotic producer).

The method described can only be used for a single indicator organism at a time and it may sometimes be useful to use a technique by which a number of replicates of the original isolation plate are prepared. These are made on the same or different test media seeded with a range of indicator organisms.

Enrichment culture

Natural environments, such as the soil, contain large numbers of different microorganisms coexisting because they all have different activities. The natural balanced community remains very heterogeneous because each species is making use of a different substrate or a different set of micro-environmental parameters. When such a diverse community is exposed to a single new substrate those strains able to utilize it most efficiently will multiply most rapidly and their numbers will form a greater proportion of the community. As they utilize the new substrate they will liberate compounds which may be substrates for other organisms and a succession of species may occur. However, if a portion of this enriched community is transferred to a fresh medium, containing the new substrate, the same organism will continue to be at an advantage and it forms an increasingly greater proportion of the community. Once the 'enriched' organism has become dominant it can be isolated in pure culture by spreading a small inoculum onto plates of solid media.

The system described above is a batch process and its success will depend on the skill and experience of the operator in choosing the right moment to transfer the enrichment culture to fresh medium, for it is possible to miss the time at which the desired organism is dominant. Such difficulties can be overcome by using a continuous culture process such as a chemostat in which fresh medium containing the novel substrate is constantly added and culture liquid constantly removed.

If it is required to isolate an organism producing a particular useful enzyme, then a substrate for this enzyme is used in the enrichment medium as the sole source of energy or carbon or nitrogen (see also section 4.1).

The desired organism is finally purified by streaking onto plates of a solid medium of the same composition.

Not only the substrate, but physical conditions such as temperature and pH may be used as a selective agent. For further detail the reader is recommended the excellent review by Veldkamp (1970).

During the search for cultures producing alkaline proteases, samples (usually 0·1 ml) from soil dilutions, previously heated (10 minutes at 80°C) to kill vegetative cells but allow the survival of spores, may be plated onto the surface of agar containing casein at pH10–12. After incubation colonies surrounded by a clear zone are subcultured.

Cultures responsible for biconversions

Rapid detection of cultures which may be useful in the bioconversion of steroids generally depend on the availability of an appropriate reaction for detecting the transformation product. This is so for the formation of Δ-7 oestrogens arising from the aromatization of steroids substituted at position 19. These give a red colour with para-nitrobenzene diazonium fluoborate (1% solution in 50% acetic acid). A layer of nutrient agar is poured into a Petri dish and covered with a thin layer of agar containing several milligrams of the steroid to be transformed as an homogeneous suspension. The plate is then inoculated by stabbing a variety of cultures to be tested. As soon as colonies appear, the reagent is sprayed over the surface of the plate and where a Δ-7 oestrogen is produced a persistent red colour appears.

It is sometimes necessary to make enrichment cultures, using a baiting method, before making isolations. Enrichment may simply consist of adding a specific bait (for example powdered cellulose) to the soil and incubating at a controlled temperature and humidity. Small samples of the soil are then incubated in shaken Erlenmeyer flasks containing a mineral medium with cellulose as the sole carbon source. After a number of subcultures it may be possible to isolate cellulolytic organisms. If the medium is slightly acidified it can be made selective for fungi and only cellulolytic fungi are recovered.

1.3   Strain development

Having discovered a microorganism producing the required compound, it is usually necessary to improve the yield. It is important to realize that microorganisms usually have regulatory mechanisms which control the amounts of metabolites synthesized so that they do not exceed the cell's requirements. To increase the yield it is necessary to induce the formation of mutants, using either physical or chemical methods, in which these regulatory mechanisms are suppressed. Such mutants may be classified into two groups; auxotrophic mutants and mutants resistant to analogues.

Cultures may be used which have multivalent mechanisms, concerted repression or feedback inhibition, and a search made for mutants which have lost the ability to synthesize one of the terminal products responsible for feedback inhibition or repression (auxotrophic mutants). Consider a situation where three terminal products, P_1, P_2 and P_3 are produced via a branched biosynthetic pathway from an intermediate A. There are two principal, distinct, regulatory mechanisms. There may be three distinct iso-enzymes, 1, 2 and 3, responsible for the first reaction in the pathway A → B and each may be inhibited or repressed by one of the three terminal products (Figure 1.7).

In the case of the multivalent or concerted regulatory mechanism, repression is only apparent if all three terminal products are present together (Figure 1.8).

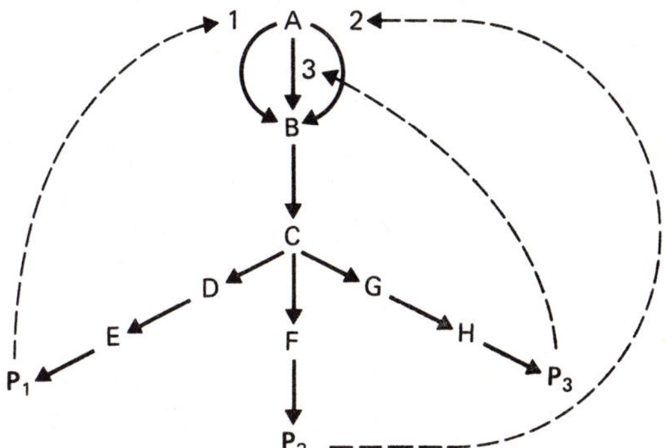

Figure 1.7   System of regulation of iso-enzymes involved in a branched biosynthetic pathway

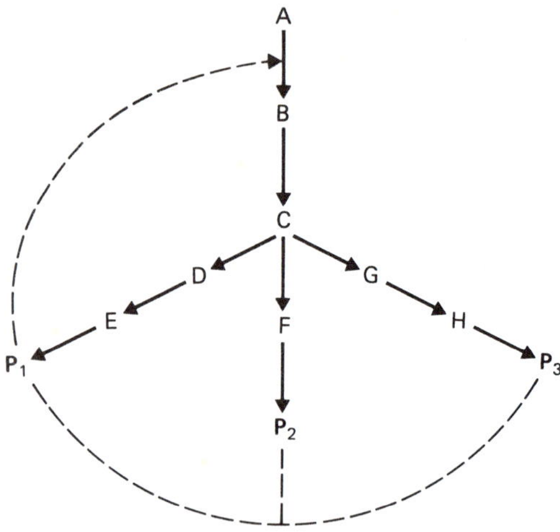

Figure 1.8   Multivalent (or concerted) regulation of a branched biosynthetic pathway

(Note: there is only a single enzyme for the reaction A → B)

It should be noted that there is sometimes a cumulative action of the three terminal products which may be represented very schematically as shown in Table 1.2.

Table 1.2   Feedback control by terminal products in a branched chain biosynthetic pathway

'End product' (in slight excess)	% inhibition or repression (multivalent)	(cumulative)
P_1	0	50
P_2	0	50
P_3	0	50
$P_1 + P_2$	0	75
$P_2 + P_3$	0	75
$P_1 + P_3$	0	75
$P_1 + P_2 + P_3$	100	87·5

A mutant of *Corynebacterium glutamicum*, which requires homoserine, excretes nearly 60 g of lysine per litre in a medium based on glucose and mineral salts. It was obtained from a wild type strain isolated from soil in 1956 by Kinoshita. The wild type does not require homoserine and does not accumulate lysine. In this wild type strain there is a common pathway to the

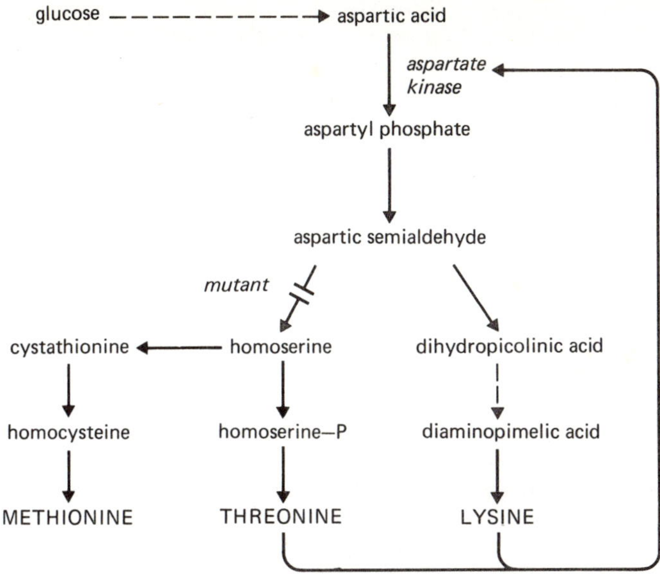

Figure 1.9   Pathway and control of lysine biosynthesis in *Corynebacterium glutamicum*

biosynthesis of lysine and threonine, for the first few reactions, which is subject to feed back inhibition by a mixture of threonine and lysine controlling the activity of aspartate kinase (Figure 1.9).

The homoserine dependent mutant no longer synthesizes threonine, feedback inhibition no longer occurs and lysine accumulates in the medium. Optimum production of lysine occurs in a medium containing 400 µg of homoserine per ml and a high concentration of biotin (20 µg/1). It can be shown that inhibition due to threonine is increased by methionine which reacts competitively with a regulatory site on aspartate kinase. The ratio of threonine to methionine also plays an important role (Table 1.3).

It should be noted that in the wild type strains of *Escherichia coli* the biosynthetic pathway to lysine is the same but its regulation is different. In this case there are three aspartate kinases, each separately controlled by either lysine, methionine or threonine. This system may be represented by Figure 1.7, P_3 being lysine. There is also feedback inhibition of dihydropicolinate synthetase by lysine.

Mutants which have lost one of the enzymes of a biosynthetic pathway may also be used for the production of an intermediate in that pathway. Consider the biosynthesis of arginine by a wild type isolate of *Corynebacterium glutamicum* shown in Figure 1.10. A mutant which cannot synthesize the

Table 1.3 Specific inhibition of lysine production by a mutant of *Corynebacterium glutamicum*

Amino acid added	Concentration	Lysine produced (μM)	(% of control)
Lysine	10^{-2} M	12·3	64
Methionine	10^{-2} M	15·9	83
Threonine	10^{-2} M	0	0
Homoserine	10^{-2} M	0	0
Lysine	10^{-3} M	28	145
Methionine	10^{-3} M	21.8	113
Threonine	10^{-3} M	0	0
Homoserine	10^{-3} M	0	0
Threonine	10^{-4} M	14.2	74
Methionine	10^{-4} M	10·6	55
Homoserine	10^{-4} M	2·4	12

enzyme acting on ornithine will accumulate that amino acid so long as just sufficient arginine is provided for growth, without enough being present to cause feedback inhibition. The best yield of ornithine is obtained when 200 μg arginine/ml and 5 μg biotin/l are present in a medium rich in glucose and ammonium salts.

Use may also be made of mutants possessing enzymes which are resistant to feedback control. For feedback inhibition to occur, a terminal product interacts with an enzyme at a regulatory site which is different from that occupied by the substrate of the enzyme. This interaction alters the configuration of the enzyme in such a way that it is no longer functional. A mutation may occur, giving rise to an enzyme with an altered regulatory site, no longer interacting with the inhibitor, and feedback inhibition no longer occurs. To obtain evidence of such a mutation it would not be sufficient to add an excess of the metabolite in question since it would then be impossible to know whether or not it was being synthesized in such a medium.

An analogue is used which, although it cannot replace the terminal product for growth of the culture, can interact with the regulatory site associated with feedback inhibition. Such an antimetabolite is often toxic thus eliminating all non-resistant mutants (Table 1.4).

In the case of threonine, the process of selection is done in two stages. First a nutrient agar medium containing the analogue of threonine (α-amino β-hydroxyvaleric acid) is poured into a Petri dish and allowed to set at an angle (Figure 1.11).

After this wedge has set, a second layer of the same medium, without

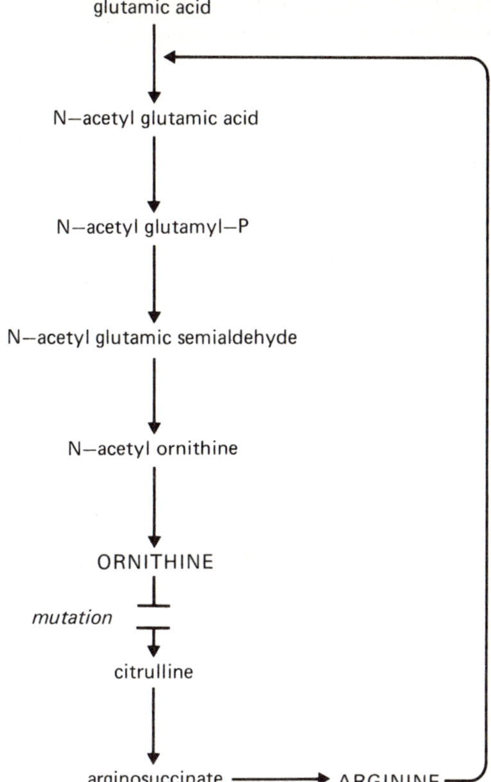

Figure 1.10   Biosynthetic pathway for arginine and its regulation in *Corynebac-terium glutamicum*

analogue, is poured onto it and allowed to set with the plate level. After a time the analogue diffuses into the upper layer giving a concentration gradient at the surface on which an inoculum, previously treated with a mutagenic agent, is spread in order to select any mutants resistant to high concentrations.

The second stage is to examine the resistant mutants for their ability to excrete threonine by inoculating them, as point cultures, onto an agar medium seeded with a threonine requiring culture. The diameter of the zone of growth of the seed organism around each colony of threonine producing mutant gives some indication of the quantity produced. Using this method a mutant of *Brevibacterium flavum* capable of accumulating up to 12·6 g of threonine per litre has been obtained.

Sometimes a mutant reverts, leading to the production of an enzyme different from that which was lost as a result of the first mutation, and resistant

Table 1.4 Mutants resistant to analogues and capable of accumulating amino acids, nucleotides, nucleosides or vitamins

Terminal product	Analogue	Culture
Phenylalanine	β-2 thienylalanine	Escherichia coli
Methionine	ethionine	Candida utilis
Methionine	norleucine	Escherichia coli
Threonine	α-amino β-hydroxy valeric acid	Escherichia coli
Inosinic acid	6-mercaptoguanine	Corynebacterium glutamicum
p-aminobenzoic acid	sulphonamide	Staphylococcus aureus
Pyridoxine	isoniazide	Saccharomyces cerevisiae

to feedback inhibition. There is an example in the threonine deaminase of a reverse mutation of a strain of *Hydrogenomonas*. Although the wild type did not require or secrete isoleucine, and the mutant was auxotrophic for isoleucine, the revertant no longer required the amino acid but also excreted it (Figure 1.12).

Numerous techniques are available for producing cultures which give increased yields of particular enzymes (Table 1.5).

In so-called constitutive mutants, the regulatory genes are so modified that the organism either produces a particular enzyme in the absence of inducers, or else one that is not susceptible to catabolic repression. Such constitutive mutants may be selected in a variety of ways. Induction can be suppressed by culturing the microbial population on a medium containing a carbon source, the metabolism of which requires the enzyme, but which does not act as an inducer. Thus phenyl β-galactoside can be used to select mutants constitutive for the production of β-galactosidase. By limiting the concentration of inducer (lactose) during continuous culture, a mutant of *Escherichia coli* has been obtained which produces β-galactosidase as 25% of its total protein.

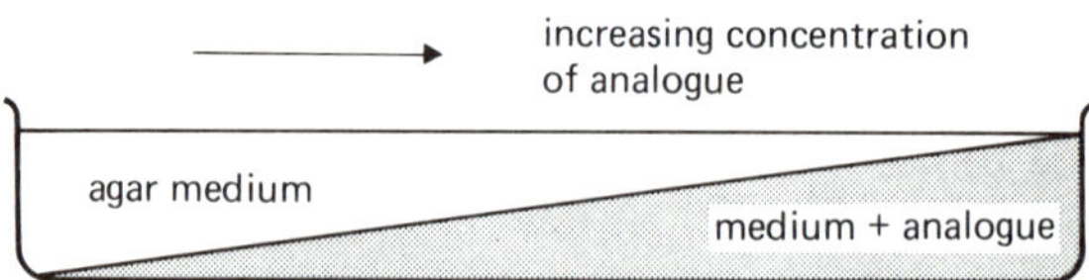

Figure 1.11   Method for obtaining a concentration gradient of an antimetabolite in agar

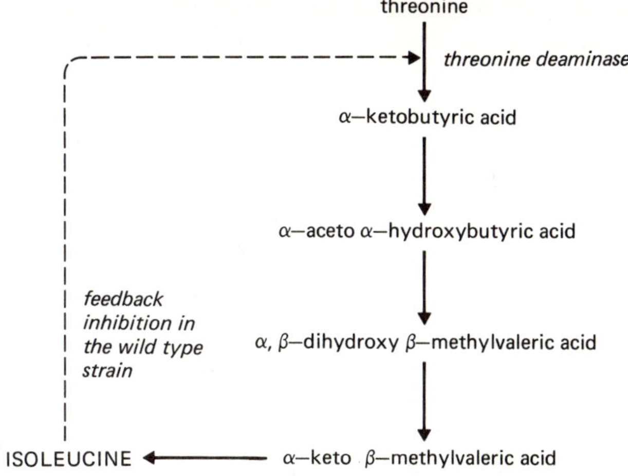

Figure 1.12   Biosynthesis of isoleucine from threonine

An alternative method consists of cultivating the organism in a cyclic manner, alternately with and without inducer, for example, in the presence of glucose then of lactose, and so on. Initially, in the presence of glucose, cells of constitutive mutants have comparable growth rates as other cells. However when glucose is replaced by lactose, constitutive mutants are favoured at first

Table 1.5   Examples of increased enzyme production by mutants

Enzyme	Microorganism	Yield (mutant/wild type)
Aspartate transcarbamylase	*Escherichia coli*	500
Cellulase	*Trichoderma viride* QM 6a	2
Dihydrofolic reductase	*Streptococcus pneumoniae*	200
Amyloglucosidase	*Aspergillus foetidus*	1·6
Protease	*Bacillus cereus*	10
α-amylase	*Aspergillus oryzae*	10
Homoserine dehydrogenase	*Escherichia coli*	3

because they do not have a lag phase. After a certain number of cycles the bacterial population will contain an increased proportion of constitutive mutants.

It is also possible to add to the medium inhibitors of the inducer, such as 2-nitrophenyl β-fucoside (in the case of β-galactosidase). If lactose is the only source of carbon in the medium only those mutants of *Escherichia coli* which produce β-galactosidase without induction can grow.

Constitutive mutants which are resistant to toxic substances, because the latter are destroyed by the enzyme for which the mutants are constitutive, are also readily isolated. When a culture of the photosynthetic bacterium, *Rhodopseudomonas spheroides*, is repeatedly exposed to 0·1 M hydrogen peroxide, 25% of the surviving mutants are constitutive for catalase.

The use of toxic antimetabolites often selects mutants producing an increased yield of enzymes involved in the biosynthesis of the metabolite concerned. Thus mutants of *Lactobacillus casei* resistant to dichloro-amethopterin produce 80 times more thymidylate synthetase than the parent culture.

Mutants resistant to the phenomenon of catabolite repression may sometimes produce useful enzymes in large quantities. They are of interest because they can be cultured on complex media in which the carbon source is cheap (as opposed to sources of carbon which do not cause catabolite repression, which are often expensive). One particular *Saccharomyces cerevisiae* mutant, for which the biosynthesis of invertase is resistant to catabolite repression, produces nearly 2% of its protein in the form of the enzyme invertase.

It is possible to obtain mutants having several copies of the gene coding for the required enzyme. A single cell of *Escherichia coli*, using lactose as its carbon source, contains about 3000 molecules of β-galactosidase which represents approximately 3% of the total protein. This is probably the maximum which can be synthesized with a single structural gene; however, it is possible to increase the number of these genes by introducing episomes or prophages during transduction. Using such methods it has been possible to obtain a culture of *Escherichia coli* having up to 25% of its protein in the form of β-galactosidase.

A quite different method of obtaining new strains is to modify the genetic information in the chromosome of the cell by transferring to it all, or part, of the DNA of another strain. Again, many techniques have been described:

Transformation

This is the transfer of a limited quantity of DNA, freely liberated into the medium by one cell and taken up by another intact cell. The extraction of DNA has to be carefully carried out and the transfer requires defined conditions. Interesting results have been obtained with antibiotic producing strains of *Streptomyces*. Transformed cultures may be produced which synthesize both streptomycin and chlortetracycline, each formed separately by two original cultures (*Streptomyces griseus* and *Streptomyces aureofaciens*).

Lysogeny

Russian authors have reported that 99% of the cells of a culture of *Streptomyces olivaceus* surviving lysis by phages had properties different from the original culture. Some of the strains thus obtained produced more antibiotic than the parent culture, extra structural genes having entered with the infecting prophage. In another example a lysogenic strain was found to synthesize ten times more transfer RNA for tyrosine than the original strain.

Transduction

It is possible to transfer the ability to synthesize streptomycin from a lysogenic strain of *Streptomyces griseus* to a non-producing strain via the prophage as an intermediary. In one experiment, the new culture produced a higher yield (5200 units/ml) than the original (4100 units/ml).

Recombination

The sexual process may be used to provide hybrid descendants (recombinants) having characters derived from both parent cultures. Although the frequency of sexual reproduction is very low in bacteria, favourable results have been obtained with cultures of the genus *Streptomyces*. Industrially useful hybrids of *Streptomyces rimosus*, combining a high yield of oxytetracycline production with low foam production in the fermenter, have been obtained in this way. Provided one only of the parent cultures is auxotrophic, a simple technique can be used to detect recombinants. If A is a wild type prototrophic culture sensitive to a bacteriostatic antibiotic, and B a resistant auxotroph, then the surface of a complete agar medium is inoculated with a mixture of A and B. When the plate is well developed a replica is made onto a minimal medium containing the antibiotic for recovering recombinants.

Heterokaryons

Amongst the filamentous fungi there is a process referred to as the parasexual cycle. The hyphae may contain two types of nuclei (each derived from a different parent) which do not normally fuse. Fusion, recombination, and fission back to the haploid state is the process called the parasexual cycle but, even without recombination, it is possible for characters of each nucleus to be expressed. When attempting to produce heterokaryons it is useful to choose two parent strains which have complementary nutritional requirements. Suppose, for example, strain 1 is able to synthesize amino acid *a* but not amino acid *b*, while strain 2 has the opposite ability. On an agar medium lacking both *a* and *b*, neither can grow alone but when inoculated together any resulting heterokaryons will form the only colonies: these may be subcultured and examined to discover any other alterations in properties which have arisen from the combination. In this way, a heterokaryotic strain of *Penicillium chrysogenum* has been obtained producing 2064 units of penicillin/ml whereas the best yield of the two parent cultures was only 1689 units/ml.

1.4   Secondary metabolites

This is a singular group of compounds which do not seem to play an immediately recognizable role in the life of the organism producing them. Characteristically they are not produced during the phase of rapid growth (trophophase) but are synthesized during the stationary phase (idiophase). They frequently have antibiotic properties (Table 1.6).

Table 1.6   Secondary metabolites characteristic of the idiophase

Actinomycin	Gramicidin
Aflatoxin	6-methylsalicylic acid
Bacitracin	Novobiocin
Chlortetracycline	Patulin
Alkaloids of ergot	Polymyxin
Erythromycin	Pyocyanin
Gibberellic acid	Streptomycin

It is not exactly understood why secondary metabolites are produced so late in the growth cycle, but it is possible that catabolic repression of certain enzymes occurs during the trophophase when carbohydrates are rapidly used. At the end of the trophophase, new enzymes involved in the biosynthesis of secondary metabolites make their appearance. In the production of penicillin (see Figure 6.1) two enzymes are produced in large quantities only after the trophophase.

The first catalyzes the activation of phenylacetic acid (which forms the side chain of benzylpenicillin), the second, called penicillin acyl transferase, catalyzes the transfer of the phenylacetyl group from the active coenzyme A ester to a derivative of 6-aminopenicillanic acid.

The yield of acyl transferase increases after 40 hours incubation and remains at a high level until 80 hours, when it begins to fall. This enzyme is found only in cultures producing penicillin and its concentration is correlated with the yield of penicillin.

During the early days of industrial production of penicillin it was appreciated that, although glucose is a good carbon source for growth, penicillin production is low in its presence. With lactose the reverse is true and in practice a medium containing a mixture of the two sugars is used. After glucose has been utilized, *Penicillium chrysogenum* becomes derepressed to utilize lactose.

This mould is only able to utilize lactose slowly and growth almost stops, whilst the biosynthesis of penicillin commences. Because lactose is expensive its use in the industrial production of penicillin has been replaced by the slow feeding of limiting concentrations of glucose during the penicillin producing stage. A similar situation occurs in the production of cephalosporin using a medium containing a mixture of glucose and sucrose: glucose is rapidly metabolized during the trophophase and sucrose is slowly metabolized during the idiophase. In the manufacture of gibberellic acid, the medium used contains two carbon sources, glycerol and lactose, and again it is possible to

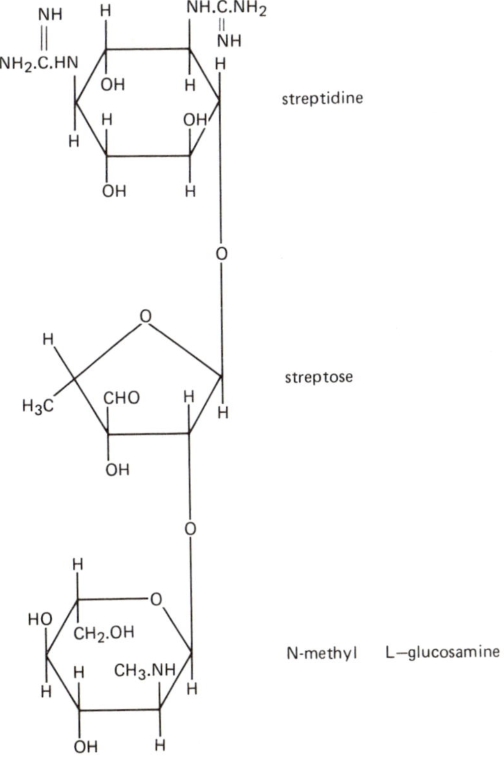

Figure 1.13   Structure of streptomycin

replace them by controlled continuous addition of glucose at low concentrations.

A good example of the effect of catabolic repression on the biosynthesis of enzymes involved in the production of antibiotics is shown by streptomycin. Cultures of *Streptomyces griseus* produce both streptomycin (Figure 1.13) and its mannoside (Figure 1.14). The latter has a low antibiotic activity (one-fifth that of streptomycin itself). At the end of the fermentation, when glucose has disappeared, an α-D-mannosidase is produced and completes the transformation of the mannoside to free streptomycin.

The production of this enzyme is subject to catabolite repression when the concentration of glucose in the medium is greater than 0·5% When too much glucose is present in the medium a mixture of the two antibiotics may be produced which would be undesirable. It has been suggested that the inhibition of the synthesis of cyclic AMP is the key factor in the phenomenon of catabolic repression.

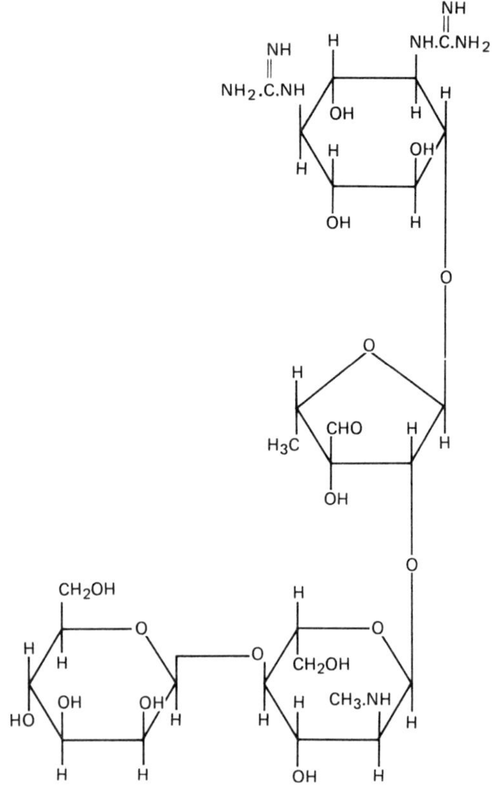

Figure 1.14   Mannosidostreptomycin

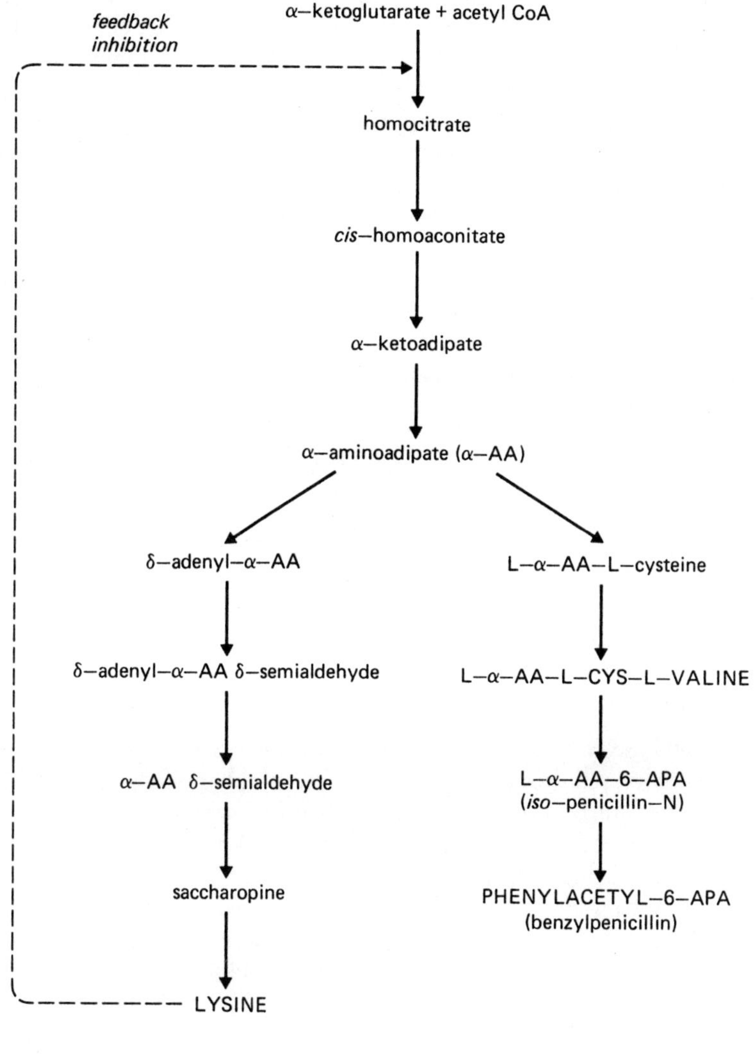

AA   = aminoadipic acid
CYS   = cysteine
VAL   = valine
6—APA = 6—aminopenicillanic acid

Figure 1.15   The branched biosynthetic pathway of lysine and penicillin in *Penicillium chrysogenum*

Feedback control also plays an important role in secondary metabolism. Thus the production of penicillin is inhibited by the presence of lysine because the two compounds share a part of a biosynthetic pathway (Figure 1.15).

The fact that the production of secondary metabolites is subject to genetic mechanisms (induction, catabolic repression and feedback control), makes it possible to obtain increased yields by mutation. In the case of penicillin, yields were increased from 5 ml/l to 10,000 mg/l during the period 1940–1970, largely by a programme of subjecting *P. chrysogenum* to successive exposure to mutagens alternating with selection to restore impaired growth vigour.

Table 1.7 illustrates the growth in yields of some principal antibiotics, using the French and Russian antibiotic industries as examples.

Table 1.7   Growth in the production of antibiotics over a period of years

Antibiotic	Initial yield at time of discovery (units/ml)	Improved yields obtained in the USSR (units/ml)	Improved yields in France, 1972 (units/ml)
Penicillin	20 (1943)	8000 (1955)	12,000–15,000
Streptomycin	50 (1945)	5000 (1955)	12,000–15,000
Chlortetracycline	200 (1948)	4000 (1959)	12,000–15,000
Oxytetracycline	400 (1950)	6000 (1959)	—
Erythromycin	100 (1955)	2000 (1961)	3000

Frequently these mutations have arisen by chance, but the use of analogues has given some useful results. For example, tryptophan is a precursor of the antibiotic pyrrolnitrin produced by a strain of *Pseudomonas aureofaciens*. By selecting strains resistant to fluorotryptophan mutants have been obtained which give yields three times those of the parent culture.

When the antibiotic is toxic to the producing strain it may be possible to increase yields by selecting strains resistant to high concentrations. This has certainly happened in the development of high yields of streptomycin. The isolation of reverse mutations has been useful in the production of mutants giving yields of chloramphenicol nine times those of the wild types strains.

1.5   Maintenance of cultures

High-yielding cultures must, of course, be carefully preserved. It is often necessary to go back to the original culture because of degeneration during the process of continuous subculturing. For example, in a study of a streptomycin producing strain of *Streptomyces griseus*, it was found that after 58 subcultures on glucose/yeast extract agar the culture no longer produced either spores or streptomycin. It should also be recognized that production by a micro-organism of large quantities of a particular metabolite, desirable though it may be to the industrial manufacturer, is biologically wasteful for that organism. Consequently, under production conditions, there may well be a tendency for selective multiplication of low-yielding but high-growth-rate variants to occur. There are many techniques for the preservation of cultures in such a way as to retain their properties. A number of these depend on maintaining the microorganisms viable but biologically inactive so that mutations are unlikely to occur during their slow vegetative growth.

Deep cultures in soil extract agar

A soft agar (3 g/l) containing soil extract (1 kg of fertile soil in 2 l water sterilized for 1 hr at 130°C and filtered) is inoculated by a central stab. After growth has occurred, the tubes are hermetically sealed and kept at 4°C.

Agar slopes under paraffin

The culture is grown on agar slopes containing peptones but no sugars. As soon as visible growth occurs, the surface is covered with sterile paraffin oil to prevent evaporation and limit the availability of oxygen. These cultures are also kept at low temperatures (4°C in a cold room or refrigerator).

Deep freezing

When a culture is frozen some of the cells die during the freezing process. During storage in the frozen state a proportion of the survivors progressively dies although the rate of death is much reduced with lower freezing tem-peratures (Table 1.8).

Table 1.8 The effect of freezing temperature on the mortality of lactic acid bacteria

Temperature of freezing	% of cells viable after	
	(24 h)	(15 days)
−7°C	49	2
−30°C	52	41

Lyophilization

When freezing is combined with desiccation the process is referred to as lyophilization. Again, the culture medium should contain a protective colloid such as milk, serum or egg-white or, in some instances, lactose. The cultures to be dried are dispensed into small tubes. These are rapidly frozen, either by immersing in liquid nitrogen or by subjection to a high vacuum (10^{-1} to 10^{-2} mm Hg) which causes freezing by loss of latent heat during partial evaporation. The frozen tubes are then subjected to a secondary vacuum drying process which removes nearly all the water, giving a light, friable deposit which can be preserved in a viable state for very long periods by sealing the tubes in a nitrogen atmosphere, or under vacuum.

Soil cultures

This is a technique applicable to sporing microorganisms (*Penicillium*, *Aspergillus*, *Bacillus* and *Streptomyces* for example). A spore suspension is first obtained which may involve the use of a special medium. A mixture of soil (20%), sand (78%) and calcium carbonate (2%) is prepared and distributed, a few grams per tube, and sterilized for 8–15 h at 130°C. After cooling, a quantity of spore suspension is added, the excess water evaporated in a desiccator under vacuum, and the tubes sealed. Spores may be kept in this way for many years without germinating.

With newly isolated cultures it is as well to use a variety of maintenance techniques: there is no *a priori* way of determining which will prove to be the best. For example, cultures of *Blakeslea trispora* do not survive freezing and should be kept at normal temperatures.

Bibliography

ALIKHANIAN, S. I., (1962), 'Induced mutagenesis in the selection of microorganisms', *Adv. Appl. Microbiol.*, **4**, 1–48.

BALL, C., (1973), 'The genetics of *Penicillium chrysogenum*', *Progr. Ind. Microbiol.*, **12**, 47–72.

BRADLEY, S. G., (1966), 'Genetics in applied microbiology', *Adv. Appl. Microbiol.*, **8**, 29–36.

DEMAIN, A. L., (1971), 'Overproduction of microbial metabolites and enzymes due to alteration of regulation', *Adv. Biochem. Engin.* **1**, 113–142.

HESSELTINE, C. W., HAYES, W. C., (1973), 'Sources and managements of microorganisms for the development of a fermentation industry', *Progr. Ind. Microbiol.*, **12**, 1–46.

MARTIN, S. M. & SKERMAN, V. B. D. (ed.), (1972), *World Directory of Collections of Cultures of Microorganisms*, Wiley-Interscience, New York and London.

MEYERS, E., SMITH, D. A., DONOVICK, R., (1968), 'Biol-autographic survey technique', *Appl. Microbiol.*, **16**, 10–12.

PIRT, S. J., (1975), *Principles of Microbe and Cell Cultivation*, Blackwell Scientific, Oxford.

RAINBOW, C. & ROSE, A. H. (ed.), (1963), *Biochemistry of Industrial Microorganisms*, Academic Press, New York and London.

REUSSER, F., (1963), 'Stability and degeneration of microbial cultures on repeated transfer', *Adv. Appl. Microbiol.*, **5**, 189–213.

SAKAGUCHI, K., UEMURA, T., KINOSHITA, S., (1971), *Biochemical and Industrial Aspects of Fermentation*, Kodansha, Tokyo.

SUOMALAINEN, H., NURMINEN, T., OURA, E., (1973), 'Aspects of cytology and metabolism of yeast', *Progr. Ind. Microbiol.*, **12**, 109–168.
VELDKAMP, H., (1970), 'Enrichment cultures of prokaryotic organisms', in *Methods in Microbiology*, Vol. 3A, Ed. J. R. Norris and D. W. Ribbons, Academic Press, London and New York.

2 Culture Media

Once a strain capable of giving a high yield of the desired product has been developed in the laboratory, it is necessary to establish a medium economically usable on an industrial scale. Liquid media are usually employed: less plant space is needed and handling problems are simplified because the liquid media and cultures are amenable to standard chemical engineering techniques and equipment.

2.1 Composition of media

The culture medium should contain all the essential elements required for growth of the microorganisms, in proportions similar to those occurring in the

Table 2.1 The main elements present in microorganisms (average composition)

Element	Physiological function	Dry weight (%)
Hydrogen	Constituent of organic compounds and water	8
Oxygen	Constituent of organic compounds and water	20
Carbon	Constituent of organic compounds	50
Nitrogen	Constituent of proteins, nucleic acids and coenzymes	14
Sulphur	Constituent of proteins and several coenzymes	1
Phosphorus	Constituent of nucleic acids, phospholipids and coenzymes	3
Magnesium	Cofactor of numerous enzyme reactions (ATP)	0·5
Manganese	Cofactor of several enzymes	0·1
Calcium	Cofactor of enzymes (proteases)	0·5
Iron	Constituent of cytochromes, non-haem proteins and cofactor of some enzymes	0·2
Cobalt	Constituent of vitamin B_{12}	0·03
Copper Zinc Molybdenum	Constituents of some enzymes	0·03

microbial cell (Table 2.1). It is customary to distinguish between the macro-nutrient elements (required in large quantities, e.g. C, H, O, N), the meso-nutrient elements (required in smaller but significant quantities, e.g. Mg, P, S) and the micronutrient (trace) elements required in only minute amounts, such as Fe, Cu, Zn and Mo.

Source of carbon

With the exception of the algae and autotrophic bacteria, which use carbon dioxide as a source of carbon, those microorganisms used in industry require organic compounds both as a source of carbon and of energy. Carbon is the most abundant element, by weight, in microbial cells, comprising approximately 50% of the biomass; the concentration of nitrogen varies from 3–15%.

The carbon source is therefore the largest ingredient of the culture medium; if it is limiting, the total biomass X is proportional to the initial concentration s of the organic source of carbon:

$$X = Ys$$

The constant of proportionality Y in this equation is termed the yield constant for the substrate and the organism, and can be determined at the beginning of the stationary phase of a batch culture. The original definition of the yield constant by Monod arises from his observation that the rate of growth of a microorganism is proportional to the rate of consumption of the rate limiting substrate:

$$\frac{dX}{dt} = -Y\frac{ds}{dt}$$

If Y were truly a constant it could be considered as the mass of biomass produced per unit mass of substrate consumed but, in fact, Y has been shown to vary with growth rate in experiments using the chemostat (Tables 2.2).

Table 2.2(a) Yield constant Y of a variety of carbon and energy sources

Medium	Culture	Number of cells/ml	Dry weight (g/l)	Y (%)
Nutrient broth with 8 g/l dry weight	*Escherichia coli*	5×10^9	1	12·5
Synthetic medium with 57 g glucose + 20 g acetic acid/l	*Serratia marcescens*	2×10^{11}	29	37·5
Mineral solution + 100 g glucose/l	*Candida utilis*	—	45	45

Table 2.2(b) Cell yield from a variety of carbon and energy sources

Organism	Substrate	Cell yield Y (g dried cells/ mole substrate)
Saccharomyces cerevisiae	glucose	20
Streptococcus faecalis	arginine	10·2
Aerobacter aerogenes	glycerol ($\mu = 0.24\,\mathrm{h}^{-1}$)	36·8
Aerobacter aerogenes	glycerol ($\mu = 0.004\,\mathrm{h}^{-1}$)	14·7
Pseudomonas $C_{12}B$	acetate	23·5

Data from Ribbons, D. W., (1970), 'Quantitative relationships between growth media constituents and cellular yields and composition, in *Methods in Microbiology* Vol. 3A, ed. J. R. Norris and D. W. Ribbons, Academic Press, London.

The two parts of Table 2.2 indicate two of the different units in which cell yields may be expressed.

Amongst microorganisms a wide range of nutritional types is found. There are versatile species which can use as carbon source a large number of compounds ranging from simple C_2 compounds (acetic acid, ethanol) to complex compounds (polysaccharides,- proteins) and even aromatic compounds. A good example is *Pseudomonas cepacia* (Table 2.3); it is worth recalling that another species of this genus (*Ps. testosteroni*) will degrade the aryl sulphonate group which terminates the alkyl chain of biodegradable detergents.

In contrast, there are microorganisms narrowly adapted to a single substrate which may well be unsuitable for the growth of others. Thus the methylobacteria *Methylomonas* and *Methylococcus* utilize only methane and methanol as sources of carbon and energy.

When the yield of ATP is known it is observed that the yield of biomass of a range of unicellular microorganisms (bacteria and yeasts) is proportional to the number of molecules of ATP formed from a number of different carbon and energy sources in defined media. It suffices to collect the biomass at the end of the growth phase and calculate the dry weight obtained as a function of ATP produced during catabolism of the energy source (Table 2.4).

Energy is obtained in essentially two ways:

(*i*) *By substrate phosphorylation.* In this case oxidation of the substrate, by loss of electrons, is accompanied by the synthesis of an 'energy-rich' phosphate which may be transferred to ADP with the formation of ATP. Thus, during glycolysis, microorganisms obtain 2 molecules ATP for every molecule of glucose consumed. This substrate level phosphorylation occurs in both

Table 2.3 Organic compounds serving as carbon sources for *Pseudomonas cepacia*

Sugars and polyols	Fatty acids	Hydroxy- and keto-acids	Amino acids	
Ribose	Acetate	Citrate	Alanine	Betaine
Xylose	Propionate	α-ketoglutarate	Serine	Sarcosine
Arabinose	Butyrate	Pyruvate	Threonine	Hippurate
Fucose	Isobutyrate	Aconitate	Aspartate	Acetamide
Rhamnose	Valerate	Citraconate	Glutamate	Nicotinate
Glucose	Isovalerate	Levulinate	Lysine	Trigonellin
Mannose	Caproate	Glycolate	Arginine	
Galactose	Heptanoate	Malate	Histidine	*Non-*
Fructose	Caprylate	Tartrate	Proline	*nitrogenous*
Sucrose	Pelargonate	Hydroxybutyrate	Tyrosine	*aromatics*
Trehalose	Caprate	Lactate	Phenylalanine	
Cellobiose		Glycerate	Tryptophan	Benzoyl-
Salicin	*Dicarboxylic*	Hydroxymethyl-	Cynurenin	formate
Gluconate	*acids*	glutarate	Cynurenate	Benzoate
2-ketogluconate				*o*- hydroxy-
Saccharate	Malonate	*Primary*	*Other*	benzoate
Mucate	Succinate	*alcohols*	*nitrogen*	*m*-hydroxy-
Mannitol	Fumarate		*compounds*	benzoate
Sorbitol	Glutarate	Ethanol		*p*-hydroxy-
Inositol	Adipate	Propanol	Anthranilate	benzoate
Adonitol	Pimelate	Butanol	Benzylamine	Phenyl-
Glycerol	Suberate		Putrescine	acetate
Butyleneglycol	Azelate		Spermine	Phenol
	Sebacate		Tryptamine	Quinate
			Butylamine	Testosterone
			Amylamine	

aerobic and anaerobic situations because the initial reactions in the aerobic degradation of glucose do not, in fact, require free oxygen.

(ii) By oxidative phosphorylation. In this case energy is conserved during the transfer of electrons along a respiratory chain. Three molecules of ATP are obtained for each pair of electrons transported from NADH to oxygen. Since there are 12 pairs of electrons available from each molecule of glucose metabolized via glycolysis followed by the Krebs cycle, 36 molecules of ATP may be obtained. Thus aerobic metabolism yields $(36 + 2)/2 = 19$ times more ATP than anaerobic. In general, aerobic microorganisms convert a much larger proportion (about 50%) of carbon substrates into biomass than do anaerobes, because they do not need to use so much for obtaining energy. It is possible to calculate the minimum quantity of a carbon substrate required in a

Table 2.4 Yield of anaerobic growth as a function of
substrate and *ATP* production

Culture	*Energy source*	*Mole ATP/ mole substrate*	*Dry weight of cells in grams* (Per mole substrate)	(Per mole ATP)
Saccharomyces cerevisiae	Glucose	2	21	10·5
Streptococcus faecalis	Glucose	2	22	11
	Arginine	1	10	10
Lactobacillus delbrueckii	Glucose	2	21	10·5
Zymomonas mobilis	Glucose	1	8·6	8·6
Propionibacterium acidi-propionici	Glucose	4	37·5	9·4
	Glycerol	2	20	10
	Lactate	1	7·6	7·6

medium to obtain a specified yield of biomass. For example, a preliminary
experiment using a medium containing an excess of the substrate might show
that it was possible to obtain a maximum of 40 g (dry weight) of bacterial cells
per litre. The biomass is assumed to contain 50% carbon and the organism to
be capable of converting, aerobically, 50% of the carbon of the substrate into
biomass carbon. The medium should therefore contain:

$$40 \times \frac{100}{50} \times \frac{50}{100} = 40 \text{ g carbon/l}$$

If the carbon source were glucose it would be necessary to add to the
medium:

$$\frac{40 \times 180}{72} = 100 \text{ g glucose/l}$$

Pure sugars such as glucose and sucrose are generally too expensive to use
industrially and it usually becomes necessary to find a cheap source of carbon,
especially as this is quantitatively the most important medium constituent.
There has thus been considerable interest in utilizing agricultural or industrial
byproducts which may not be otherwise usable and may be stable in price.
Moreover, a fermentation industry may choose between several primary
materials depending on their cost, and contribute to regulating an essentially
fluctuating market by using surpluses.

Molasses is a byproduct of the cane and beet sugar industry. Beet molasses
has the following composition (Table 2.5).

Table 2.5 Composition of beet molasses as % dry weight

		Approximate ash composition (%)	
Water	16·5		
Dry weight	81–83·5		
Sucrose	48–50	K_2O	45
Raffinose	1	Na_2O	15
Glucose + fructose	1	P_2O_5	0·7
Nitrogen compounds	12–13	CaO	3
Glutamic acid	3·5	MgO	1·8
Other amino acids	5·5	SO_4^{2-}	6·3
(asparagine, aspartic		SiO_2	1·1
acid, alanine, glycine)		Cl^-	18
Betaine	3·25–4·25	Fe_2O_3	0·2
		Al_2O_3	0·8
Ash	11–12		

Vitamins such as biotin, pyridoxine, thiamine, pantothenic acid and inositol are also present. Cane molasses is particularly rich in biotin, pantothenic acid, thiamine, phosphorus and sulphur. It contains less organic nitrogen than beet molasses because there is no betaine, but this substance is not assimilated by yeasts. It contains 62% sugar of which 32% is sucrose, 14% glucose and 16% fructose.

Milk whey, obtained as a byproduct of cheese making, has an equally well balanced composition (Table 2.6). However it is often more profitably sold for use as a pig food.

Table 2.6 Composition of milk whey (g/l)

Lactose	45–50
Protein	7–9
Soluble nitrogenous compounds	1·5
Lipids	1–2
Mineral salts	6–8
Dry weight	63–70

The sulphite waste liquors from the paper pulp industry contain sugars derived from the hydrolysis of hemicelluloses in wood. Their composition varies according to the wood used. Angiosperms give sulphite liquors containing 3% sugars of which some 70% are pentoses (principally xylose), whereas gymnosperms produce liquors containing about 2% sugars, some 75% of which are hexoses (principally mannose). Acid hydrolysis of wood cellulose itself may give 65–85% fermentable sugars. Cellulose usually has to be hydrolyzed before it can be used as a substrate but, using cellulolytic microorganisms, it is possible to obtain microbial protein directly from untreated cellulosic wastes; filamentous moulds (*Trichoderma viride*) and bacteria (*Cellulomonas* species) are both being studied in this context. There are several methods of acid hydrolysis which were developed mainly during the second world war. Sulphuric acid of about 0.5% concentration is usually used at 150–185°C. In a continuous process it is possible to obtain from sawdust a syrup containing 4–5% reducing sugars (a mixture of glucose and pentoses) with an overall yield of 45–55%. This syrup may be concentrated to give a kind of wood molasses (Table 2.7).

Table 2.7 Composition of wood molasses (% by weight)

Solids	60–52
Reducing sugars (as glucose)	48–50
Other carbohydrates	0·5–1·5
Non-carbohydrate organic compounds	6–8
Ash	2–3
Nitrogen	0·065
Volatile organic acids	1–2

In wood molasses from conifers the sugar mixture comprises about 85% hexoses and 15% pentoses and from broad leaf trees, about 65% hexoses and 35% pentoses. Pentose sugars are readily utilized by *Candida utilis*. Recently it has been shown that powdered and heated cellulose is susceptible to enzymic hydrolysis. A culture filtrate of *Trichoderma viride* may be used as the enzyme preparation. Using a continuous system, involving ultrafiltration membranes, all of the cellulose is hydrolyzed to glucose which passes through the membrane.

Starch (cereals, potato) may also be a cheap source of carbon after hydrolysis using plant enzymes (malt) or microbial amlyases (bacterial or fungal). A continuous process (the Symba process) developed in Sweden for biomass production utilized the yeast *Endomycopsis fibuliger* to hydrolyze starch to fermentable sugars; *Candida utilis* was subsequently grown on these sugars.

Gas oil (crudely fractionated petroleum oil) contains 10–25% of the C_{10}–C_{18} paraffin hydrocarbons which are those most easily usable by microorganisms. Hydrocarbons having the general formula C_nH_{2n+2} may have linear (normal) or branch chain-molecules (iso-): the latter are degraded very slowly, if at all, by microorganisms. In a petroleum fraction such as gas oil there may be other hydrocarbons (cyclic paraffins and aromatic compounds) which are usually less readily metabolized. It is possible to produce microbial biomass by growth on a petroleum fraction and, at the same time, make use of the organism's substrate specificity. Thus, the yeast *Candida lipolytica* metabolizes normal paraffins only; after its growth on a petroleum oil fraction, a material enriched in isoparaffins is left. Since the isoparaffins have more desirable properties as fuels, this is a useful contribution to the refining process. Alternatively it is possible to isolate the normal paraffins of gas oil by various physical methods, such as molecular seiving, and use the purified material as substrate for biomass production.

Hydrocarbons have now become important as carbon and energy sources for the growth of microorganisms, mainly in the production of feed protein (Single Cell Protein, S.C.P.); a number of bacteria and fungi are capable of growth on them. Strains from the genera *Pseudomonas, Bacillus, Acinetobacter* (*A. calcoaceticus*), *Corynebacterium, Mycobacterium* (*M. smegmatis*), *Nocardia* and from the family Enterobacteriaceae are representative of the bacteria. Some filamentous moulds can assimilate both normal paraffins (*n*-alkanes) and olefins (*n*-alkenes); *Cladosporium resinae* assimilates fuel oils and may be responsible, in part, for the corrosion of fuel tanks. Among the yeasts, Cryptococcaceae seem best able to use hydrocarbons; perhaps their wall structure is resistant to the solvent properties of hydrocarbons. Species of the genus *Candida*, such as *C. lipolytica*, have given the best results from an economic point of view. Studies have been made of the relationship of the substrate specificity amongst strains of the genus *Candida* (Figure 2.1).

It can be seen that these yeasts do not assimilate alkanes shorter than C_9 and that the alkanes C_{11} and C_{18} appear to be most widely assimilated. It is possible that alkanes of low molecular weight are toxic because, being capable of dissolving lipids, they destroy the phospholipid micelles of cell membranes. Some strains of *Pseudomonas* can grow on hexane, heptane and octane as

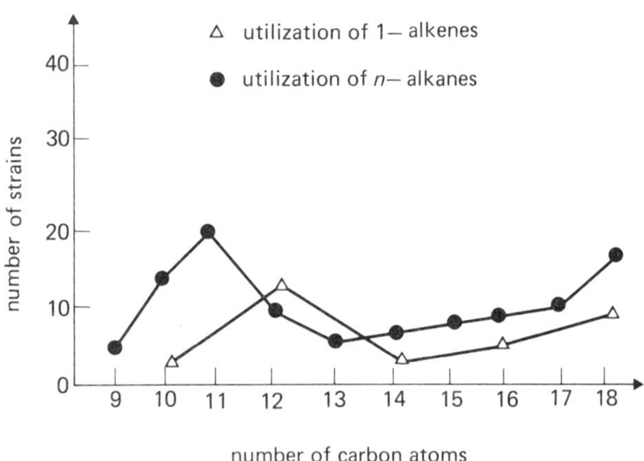

Figure 2.1   Relation between the number of strains of the genus *Candida* metabolizing hydrocarbons of different chain length

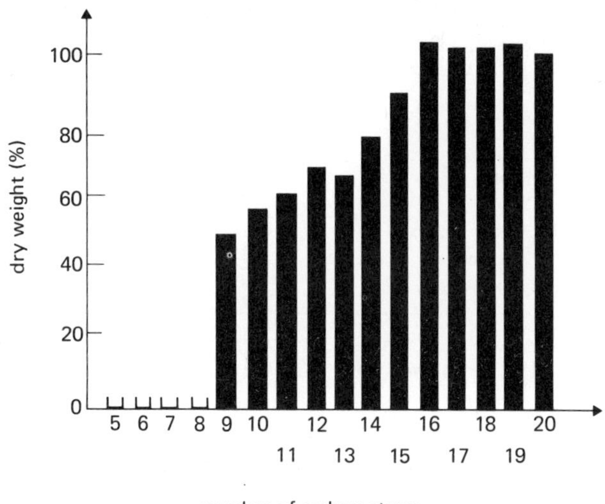

Figure 2.2   Growth of a strain of *Candida lipolytica* 4.1 (expressed as dry weight after 70 h culture) as a function of the number of carbon atoms in the molecule of alkane utilized (Growth on hexadecane is taken to be 100%)

sources of carbon and energy; the biomass obtained is a function of the number of carbon atoms in the *n*-alkanes the optimum being between 16 and 20 (Figure 2.2).

In mixtures of alkanes, those having the shorter chain length are degraded more rapidly. Hydrocarbons have both advantages and disadvantages as growth substrates, when compared with the carbohydrates. Their low solubility in water (Table 2.8) makes it possible to recover, by centrifugation, any proportion or fraction which has not been used for microbial growth. On the other hand, this low solubility causes problems with the transport of substrate from the medium to the cell. It is likely that alkanes enter the cell both as dissolved and dispersed hydrocarbon. Cells of yeasts growing on alkanes have a characteristic morphology: the cytoplasmic membrane grows thicker and shows numerous invaginations. Transport through the cell wall is presumably passive and only transport across the cytoplasmic membrane is an active process requiring metabolic energy. Because the enzymes involved in the first stages of oxidation are localized on the outside of the cytoplasmic membrane it is believed that active transport is of the fatty acids. Catabolism of normal paraffins involves an oxygenase, there being evidence from experiments using labelled oxygen, that molecular oxygen is directly incorporated into the molecule. Oxygen may react with C_1 or C_2, thus a culture of *Methylomonas methanica* oxidizes propane to give a mixture of propionic acid and acetone.

Table 2.8 Solubility of n-alkanes in water at 25°C

Alkane	Molar concentration in saturated solution
Hexane	$1\cdot1 \times 10^{-4}$
Octane	$5\cdot8 \times 10^{-6}$
Decane	$3\cdot1 \times 10^{-7}$
Dodecane	$1\cdot7 \times 10^{-8}$
Tetradecane	$9\cdot8 \times 10^{-10}$

Figure 2.3 indicates a possible pathway for the oxidation of normal paraffins.

Further degradation of the fatty acid may occur by β-oxidation or possibly ω-oxidation with the formation of a dicarboxylic acid (Figure 2.4).

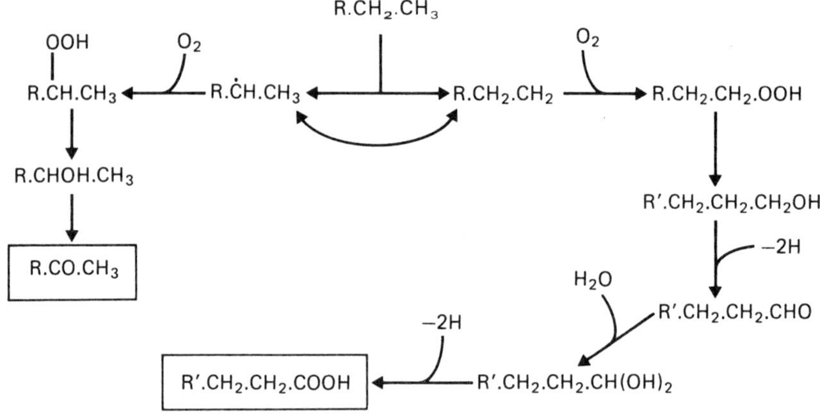

Figure 2.3   Possible pathway for the oxidation of normal paraffins

A number of 'fermentation' products may be obtained from hydrocarbons (Table 2.9). A major inconvenience in the use of hydrocarbons as a source of carbon is that considerable quantities of oxygen are required for, unlike other substrates, they contain no oxygen. Considering an imaginary example of the formation of microbial biomass (global composition $C_5H_9NO_4$) it is possible to formulate the following reactions for two different substrates:

Glucose

$$C_6H_{12}O_6 + 2{\cdot}1O_2 + NH_3 \rightarrow 1{\cdot}2C_5H_9NO_4 + 5{\cdot}4H_2O$$

(The molecular yield is 120% or 98% by weight.)

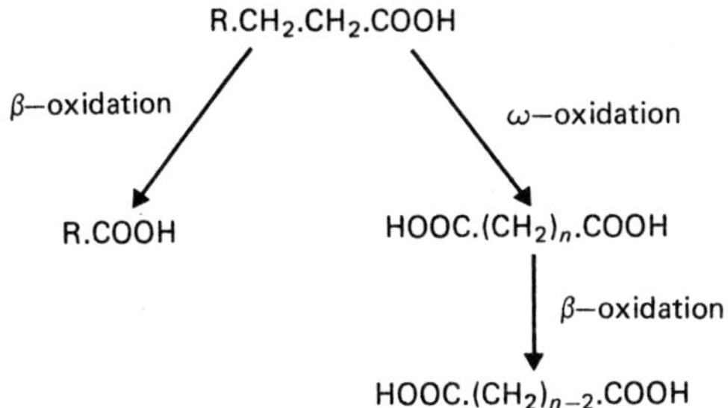

Figure 2.4   Further degradation of fatty acids following oxidation of normal paraffins

Table 2.9   Products obtained from hydrocarbons using microorganisms

Product	Culture	Substrate
Glutamic acid	*Corynebacterium hydrocarboclastus*	*n*-alkanes
Cytochrome C	*Candida albicans*	*n*-hexadecane
Carotenes	*Mycobacterium smegmatis*	*n*-alkanes
Coenzyme Q	*Candida tropicalis*	*n*-alkanes
Salicylic acid	*Pseudomonas aeruginosa*	naphthalene
Lactic acid	*Arthrobacter globiformis*	1,2-propandiol
L-tryptophan	*Bacterium* sp.	hydrocarbons
L-citrulline	*Corynebacterium* sp.	*n*-alkanes
Vitamins (B group)	*Candida* sp.	*n*-alkanes
Pyrrolnitrin	*Pseudomonas cepacia*	toluene
Cunic acid	*Pseudomonas acidovorans*	*p*-cymene
Toluic acid	*Pseudomonas aeruginosa*	*p*-xylene
Trans-4-ethyl cyclohexanol	*Alcaligenes faecalis*	ethyl cyclohexane

Hexadecane

$$C_{16}H_{34} + 8{\cdot}45O_2 + NH_3 \rightarrow 3{\cdot}2C_5H_9NO_4 + 4{\cdot}1H_2O$$

(The molecular yield is 320% or 208% by weight but requires four times as much oxygen.)

In the production of glutamic acid, it has been shown experimentally that when hexadecane was used instead of glucose, three times more oxygen was required and three and a half times more heat was produced (10·35 kcal/g glutamic acid formed from hexadecane compared with 2·78 kcal/g from glucose).

Aliphatic hydrocarbons are not the only ones available for metabolism. A number of commercially available cyclic hydrocarbons can be utilized (benzene, toluene, styrene, naphthalene and cyclohexane). The metabolism of

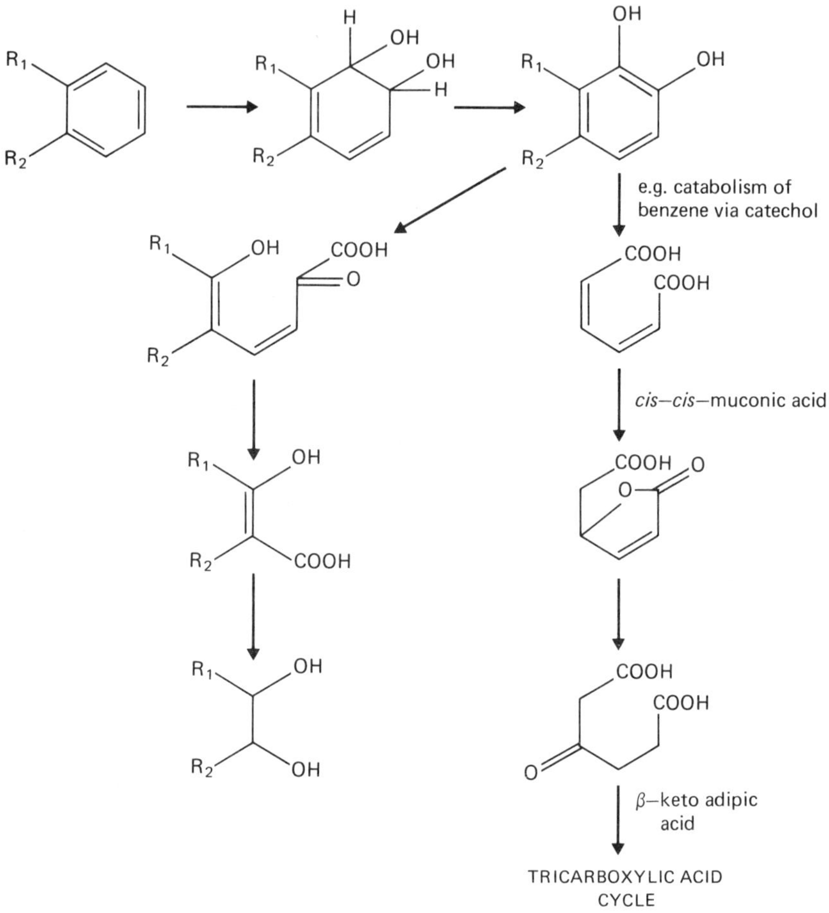

Figure 2.5   A simplified scheme for the opening of the nucleus of aromatic hydrocarbons

aromatic hydrocarbons becomes complex once the ring is opened (Figure 2.5).

The aromatic ring is first hydroxylated then dehydrogenated to a dihydroxy compound which is opened, in one of several possible ways, by an oxygenase.

A number of useful products (fatty acids, amino acids, citric acid, vitamins and carotenoid pigments, Table 2.9) may be obtained from hydrocarbons as well as from conventional carbon sources such as carbohydrates. However, many compounds, including most antibiotics and the end products of anaerobic fermentation, cannot at present be produced from hydrocarbons.

Cyclic hydrocarbons and short chain paraffins are usually toxic at high concentrations so it is necessary to develop special techniques for maintaining low concentrations. Using such techniques it is possible to obtain products not produced from other substrates, such as benzoic acid from toluene.

Source of nitrogen

All microorganisms capable of using inorganic nitrogen can assimilate ammonia or ammonium salts. Nitrates must be reduced to the equivalent of ammonium, and yeasts, as a group, are not characteristically capable of doing this. The growth of microorganisms, using ammonium salts as nitrogen source, is frequently increased by the addition of amino acids or purine and pyrimidine bases to the medium. This is usually because the rate of synthesis of one or more of these compounds is limiting the growth rate. In the antibiotic industry, corn steep liquor has long been employed as a nitrogenous medium constituent; it is a byproduct of maize starch production, obtained by concentrating the liquors from the initial steeping of the grain. It contains $7–8\%$ nitrogen, of which more than a third is in the form of amino acids, and is rich in nucleosides and vitamins.

Some microorganisms, such as the lactic acid bacteria, require certain amino acids which they are incapable of synthesizing. Thus various relatively expensive sources of amino-nitrogen, such as soya flour, fish meal and ground nut meal have to be used. Cooked cereals mashed with barley malt usually contain ample sugars and amino acids for the growth of microorganisms because of the amylases and proteases produced during germination of the barley. Media based on milk whey are also rich in amino acids as a result of the lactalbumin and lactoglobulin content. Sometimes it may be possible to utilize wastes from abattoirs and certainly, collagen, which is rich in glycine, gives improved yields of riboflavin.

The question of nitrogen nutrition is very complex. Peptides frequently stimulate growth and indeed enzymic hydrolysates of proteins are often of greater value in the growth of certain microorganisms than the equivalent mixture of free amino acids. For example, *Streptococcus faecalis* shows only feeble growth in a medium containing arginine and other amino acids known to be required. If, however, the mixture of amino acids is replaced by peptides containing arginine, growth is abundant. Cell extracts of *Streptococcus faecalis* are rich in the enzyme arginine dehydrogenase, and it has been suggested that, whereas the peptides pass the cytoplasmic membrane and are then utilized for the synthesis of proteins, free amino acids may be metabolized making them unavailable for protein synthesis (e.g. arginine is dehydrogenated).

The so-called streptogenins, growth factors for various streptococci, are low molecular weight peptides found in tryptic hydrolysates of proteins. The exact structure of these peptides has not been completely elucidated, but synthetic peptides, such as seryl-glycyl-glutamic acid, have been found to show the same activity.

Growth factors

These are those essential metabolites which certain microorganisms are
incapable of synthesizing for themselves. Some microorganisms require no
growth factors whatever; for example, a number of species of bacteria grow
satisfactorily on simple glucose-ammonium-salts mixtures. When growth
factors are needed, these are usually B-group vitamins or related compounds,
certain amino acids or, sometimes, fatty acids. When a particular growth
factor is required only in low concentration the microorganism is said to be
auxotrophic for that factor. The growth of auxotrophs is strictly proportional
to the concentration of the factor, within a particular range, making it possible
to use them for the biological assay of compounds such as vitamin B_{12}. Such
assays can be very sensitive; for example, suitable strains of the lactic acid
bacterium *Lactobacillus arabinosus* respond to the presence of 10^{-7} g of
biotin/l.

Many of these growth factors are involved in enzyme reactions because they
form a part of the structure of various coenzymes (Table 2.10).

Table 2.10 Relations between vitamins of the B group (growth factors)
and certain coenzymes

Vitamin	*Coenzymes*	*Enzyme activity*
Nicotinic acid	NAD & NADP	Dehydrogenations
Riboflavin (Vitamin B_2)	Flavin nucleotides	Dehydrogenations and electron transport
Thiamine (Vitamin B_1)	Cocarboxylase	Decarboxylations
Pyridoxine (Vitamin B_6)	Pyridoxal phosphate	Metabolism of amino acids (trans- and deamination, decarboxylation)
Pantothenic acid	Coenzyme A	Oxidation of keto acids, fatty acid metabolism.
Folic acid	Tetrahydrofolic acid	Transfer of C_1 compounds
Biotin	Enzymes with biotin as prosthetic group	CO_2 fixation and transfer of the carboxyl group
Cobamide (Vitamin B_{12})	Cobamide coenzymes	Molecular rearrangements

Growth factors are usually present in the materials used as a source of
carbon and of nitrogen in the formulation of a medium. It may sometimes be
necessary to supplement a medium with a specific source of growth factors,
such as yeast extract, liver extract or distillation residues to increase yields (e.g.
in the production of riboflavin by fermentation).

It is possible to have too high a concentration of a growth factor. One example has already been discussed during the study of auxotrophic mutants used for amino acid production. Thus, to obtain lysine by fermentation, the quantity of threonine or methionine (which are growth factors for the lysine producing culture) should not exceed 200 mg/l. Again, some molasses cannot be used as a carbon source for glutamic acid production because it contains too high a concentration of biotin. The yield of glutamic acid is a function of the concentration of biotin in the medium (Figure 2.6), the optimum being about 2–5 μg/l (corresponding to about 0·5 μg biotin per gram of cells).

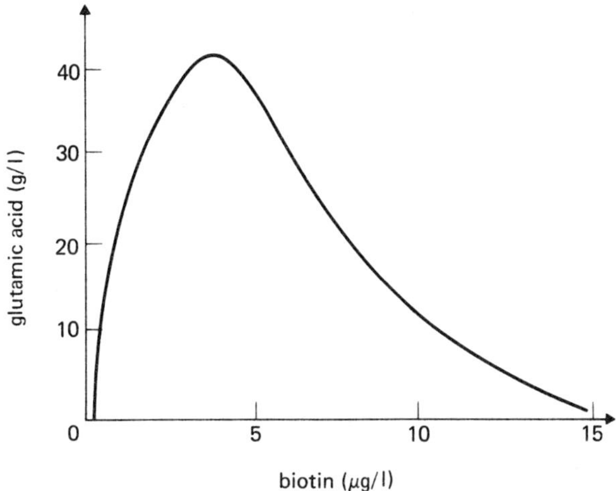

Figure 2.6   Yield of glutamic acid as a function of the concentration of biotin in the medium

The mechanism by which biotin affects glutamic acid production is unclear, but several observations suggest that it may be concerned with an effect on cell permeability. Table 2.11 shows the yields of glutamic acid produced by *Corynebacterium glutamicum* in media rich (25 μg/l) or poor (2·5 μg/l) in biotin. Bacteria grown in a medium rich in biotin do not excrete glutamic acid when resuspended in a biotin deficient medium, but under a reverse of these conditions, glutamic acid release does occur. Moreover the addition of penicillin (2–4 μg/l), or detergents such as Tween 40 or Tween 60, facilitate excretion of glutamic acid by this same organism even in the presence of high concentrations of biotin; the morphology of the cells is modified and they resemble more those grown in a medium containing a low concentration of biotin. (Tween 40 = polyoxyethylene sorbitan monopalmitate; Tween 60 = polyoxyethylene sorbitan monostearate).

Table 2.11   The effect of biotin concentration on the levels of amino acid present in the cells of *Corynebacterium glutamicum* and in the culture medium

Amino acids (μmoles/g dry weight cells)	Medium rich in biotin		Medium poor in biotin	
	(medium)	(cells)	(medium)	(cells)
Aspartic acid	21	0	73	0
Glutamic acid	85	49	15,420	3
Alanine	254	32	1,019	2
Valine	63	5	91	0
Leucine	7	2	0	1

Studies with an oleate-requiring mutant of *Brevibacterium thiogentalis*, have demonstrated the involvement of biotin in determining the composition of cell lipids. The wild type strain excreted glutamic acid when growing in a medium containing low concentrations of biotin or oleic acid. The mutant was independent of biotin for growth but required oleic acid, due to loss of an enzyme in the biosynthetic pathway of oleic acid subsequent to those reactions involving biotin. Provided the medium contained small amounts of oleic acid, it excreted glutamic acid irrespective of biotin concentration. Furthermore the oleic acid content of cells of the mutant was a function of the concentration of oleic acid in the medium, whereas in the wild type grown without oleate, it was a function of the concentration of biotin in the medium.

Mineral salts

Functions of some of the mesonutrient and micronutrient mineral elements have already been given in Table 2.1. It is difficult to determine the exact mineral salts requirement of a microorganism but they do affect the yields of some products. There are two general forms of approach:

(i) *Analytical*

The concentration of mineral elements in the ash, which represents about 5% of the microbial biomass, is measured and a medium is formulated to contain these elements in similar relative proportions. Generally, some 60% of this ash is in the form of phosphorus, 20% as sulphur; the rest is made up of a variety of metals many of which play an important catalytic role and may have to be provided in crucial concentrations. A defect of the analytical approach is that, because microbial assimilation is not very selective, the composition of the ash tends to vary according to the composition of the culture medium.

(ii) Synthetic

By trial and error, a medium which will give good growth and which is composed of chemically defined constituents is devised (a 'synthetic' medium). Various mineral elements are then successively omitted to determine those which are essential. Difficulties arise from the presence of impurities in even analytical grade reagents. Thus an iron salt, though appearing to be spectroscopically pure, may still contain up to 0.1% of cobalt: the concentration of cobalt for optimum growth is, in fact, generally of the order of 10^{-12} g/ml. It is possible to add measured amounts of chelating agents (e.g. ethylenediaminetetracetic acid, EDTA), which bind very strongly to metal ions making them unavailable, to a medium to create a total deficiency; the minimum quantity of a metal ion required to abolish this deficiency is then determined.

Phosphorus and sulphur requirements are generally satisfied by providing phosphates and sulphates in relatively large quantities, of the order of 0.5 g/l. Other inorganic elements are required in much smaller quantities, and are referred to as oligonutrients, but a precise concentration range is sometimes essential. Iron, for example, is primarily involved in the constitution of transport molecules based on haem, such as the cytochromes (Figure 2.7). It is

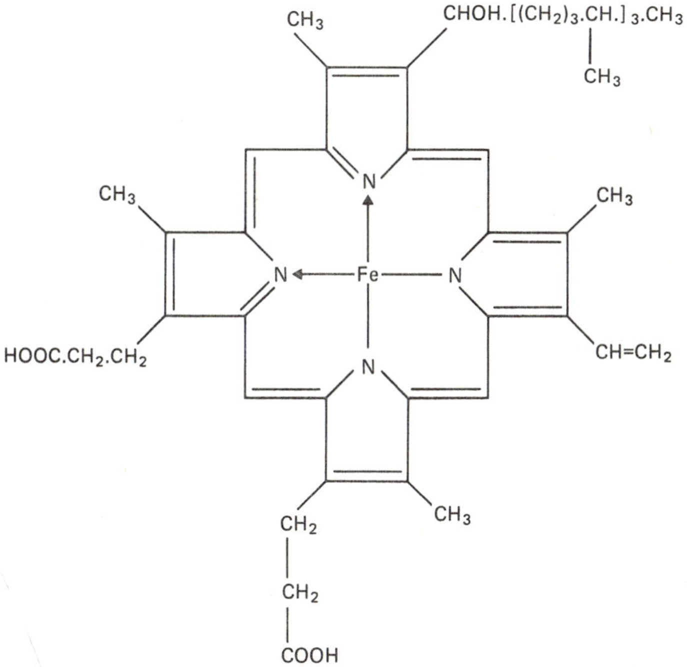

Figure 2.7　Constitution of the haem of cytochromes a and a_1

required at a concentration between 0·03–0·3 g/l in media for aerobic organisms, but the precise iron concentration influences the properties of the microorganisms grown and this may sometimes be very important in industrial fermentations. *Streptomyces griseus* requires five times more iron to produce streptomycin than it requires for growth; *Corynebacterium diphtheriae* produces its toxin only if the concentration of iron in the medium is less than 0·8 ppm; *Candida guilliermondii* synthesizes riboflavin only in media containing less than 0·01 μg/l; the mould *Aspergillus niger* produces citric acid from glucose (or sucrose) only if the medium is deficient in iron (in this last example, it has been suggested that citric acid accumulates only if the organism lacks one of the constituents required for activating the enzymes *cis*-aconitase and isocitrate dehydrogenase of the tricarboxylic acid cycle; when the carbon source is molasses, it is necessary to precipitate iron by adding ferrocyanide to obtain a high citric acid yield).

It is necessary to consider, also, possible interactions between ions. Cobalt, for example, is generally required in concentrations of about 5 ppm because it is an essential component of active vitamin B_{12}. A concentration of 10 ppm may exert a toxic effect, but this can be eliminated by the presence of magnesium. Specific requirements for oligonutrients depend both on the organisms employed and the product which is sought. Optimum growth of *Lactobacillus plantarum* requires 1 mg/l of manganese and only 0·1 mg/l is required by *Streptococcus faecalis*. To obtain high yields of inosine monophosphate (IMP) by fermentation using *Brevibacterium ammoniagenes*, the concentration of manganese must be less than 10 μg/l.

A generally useful balanced mineral medium might have the following composition per litre of twice distilled water:

NH_4Cl	1·0 g
K_2HPO_4	0·5 g
$MgSO_4.7H_2O$	0·5 g
$CaCl_2$	10 mg
$FeSO_4.7H_2O$	1 mg

These salts, even when analytically pure, contain sufficient impurities to obviate the need for deliberate additions of such elements as Cu, Zn, Mn, Mo, Co. For example, when a hypothetical salt of molecular weight 100, containing 1 in 10,000 impurities, is added to a medium at the rate of 1 g/l, the number of molecules of the salt will be

$$6 \times 10^{23}/10^2 = 6 \times 10^{21}/l$$

A bacterial culture containing 10^{12} organisms/litre would thus have available

$$6 \times 10^{21}/6 \times 10^{12} \times 10^4 = 6 \times 10^5$$

molecules of impurities available for each cell.

2.2   Preparation of media

Certain media constituents may require preliminary treatment: starch or cellulose may need to be hydrolyzed and, in the case of sulphite liquors, it is necessary to eliminate sulphur dioxide which is inhibitory to most micro-organisms. Once these preliminary processes are completed the various constituents are dissolved or suspended in water and sterilized.

2.3   Sterilization of media and plant

The purpose of plant and media sterilization is to prevent contamination of the end-product and to ensure that all of the nutrients are available for growth of the desired organism. Absolute or virtual sterility is easily obtained on a laboratory scale by autoclaving for short periods. On an industrial scale, the difficulties arise from the large quantities involved, a single surviving spore in a batch of medium being theoretically capable of causing contamination, from the damage to medium constituents caused by lengthy heat treatment and from the additional need for sterilizing bulky and complex chemical engineering equipment which will contain the growing cultures. The solution may be a compromise; the culture is worked up stage-wise, in sterile conditions to provide a large, pure inoculum in the exponential growth-phase. This is then used to inoculate the production batch of medium (clean but not necessarily sterile) in the hope that it will outgrow any contaminants present. Selective growth conditions can also hinder contamination, for example the medium may be acidic (as in the culture of yeast), or a thermophilic organism may be used; in antibiotic production the antibiotic itself confers selective protection once its formation begins.

In other processes rigorous asepsis is essential if high yields are to be obtained (Table 2.12).

Table 2.12   Production of tetracycline by *Streptomyces aureofaciens*

Fermentation conditions	Yield of tetracycline (mg/l)
Total asepsis	1328
Contaminated with the yeast *Candida utilis*	794

Because of the very large volumes of liquids to be sterilized ionizing radiation or ultrasonic treatment are impracticable for sterilization. The empty fermenters may be sterilized with gaseous chemicals such as ethylene oxide or the vapours of liquids such as β-propiolactone, but high pressure steam is almost universally used because it is more practical, less expensive and less hazardous. Steam is used both for sterilizing media in batch cookers and for continuous sterilization employing either steam injection and flash cooling or plate heat exchangers.

The sensitivity of microorganisms to heat at a constant temperature

The destruction of microorganisms by heat can be demonstrated by maintaining a suspension of bacteria or spores at a constant temperature and counting the survivors over a period of time. If the time of heating is plotted on the abscissa, and the logarithm of the number of survivors on the ordinate, then a straight line relationship is obtained (Figure 2.8). Such survival curves at constant temperature may be referred to as the Thermal Death Time Curves. From the straight line relationship it is possible to deduce that the rate of thermal destruction is directly proportional to the number of organisms N present at time t.

$$\frac{dN}{dt} = -kN \quad \text{or} \quad \frac{dN}{N} = -k\,dt \tag{2.1}$$

$$\ln\frac{N_2}{N_1} = -k(t_2 - t_1) \tag{2.2}$$

$$\ln\frac{N_1}{N_2} = k(t_2 - t_1) \tag{2.3}$$

$$\log_{10}\frac{N_1}{N_2} = \frac{k}{2\cdot3}(t_2 - t_1) = \frac{(t_2 - t_1)}{2\cdot3/k} \tag{2.4}$$

The constant $2\cdot3/k$ is called D (the decimal reduction time) and if N_0 represents the number of viable cells at the start and N the number of survivors at time t, then the equation may be written simply as:

$$\log_{10}\frac{N_0}{N} = \frac{t}{D} \tag{2.5}$$

For the time $t = D$, $\log_{10} N_0/N = 1$ and $N_0/N = 10$ from which it can be seen that D represents the time required to reduce the population size by a factor of ten (Figure 2.8).

The values of both k and D have been determined for a number of bacterial spores at a temperature of $250°F = 121°C$ (Table 2.13). In fact, it is the heat

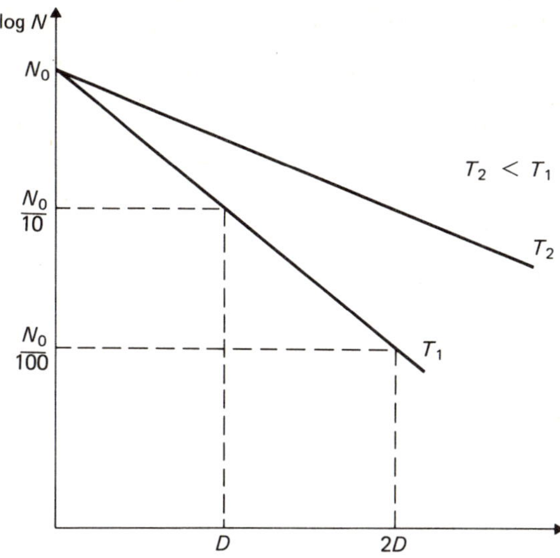

Figure 2.8   Survival curve of bacteria at two temperatures

resistant spores of bacteria which give particular problems during sterilization, the vegetative cells and the spores of fungi being fairly sensitive to heat treatment.

Table 2.13   Values of k and D for various bacterial spores at 121°C

Culture	k (min^{-1})	D (min)
Bacillus subtilis FS 5230	2·6–3·8	0·6–0·9
Bacillus stearothermophilus FS 1518	0·77	3
Clostridium sporogenes PA 3679	1·8	1·8

In practice, what is required is a curve for the heat resistance of a contaminant microbial population derived from media constituents, such as corn steep liquor, and from the environment; values of k and D are thus determined experimentally for any particular situation rather than extrapolating from values obtained with pure cultures.

Sensitivity of microorganisms to heat as a function of temperature

From Figure 2.8 above showing survival curves for the same suspension of spores at two temperatures T_1 and T_2 $(T_2 > T_1)$, it can be seen that increasing temperature brings about a reduction in the constant D.

It can be shown experimentally that the rate of change of D with temperature is proportional to D.

$$\frac{\mathrm{d}D}{\mathrm{d}T} = -k'D \quad \text{or} \quad \frac{\mathrm{d}D}{D} = -k'\,\mathrm{d}T \tag{2.6}$$

$$\ln\frac{D_2}{D_1} = -k'(T_2 - T_1) \quad \text{or} \quad \ln\frac{D_1}{D_2} = k'(T_2 - T_1) \tag{2.7}$$

$$\log_{10}\frac{D_1}{D_2} = \frac{k'}{2\cdot3}(T_2 - T_1) = \frac{(T_2 - T_1)}{2\cdot3/k'} \tag{2.8}$$

If the constant $2\cdot3/k' = Z$ then the equation may be written simply as

$$\log_{10}\left(\frac{D_1}{D_2}\right) = \frac{(T_2 - T_1)}{Z} \tag{2.9}$$

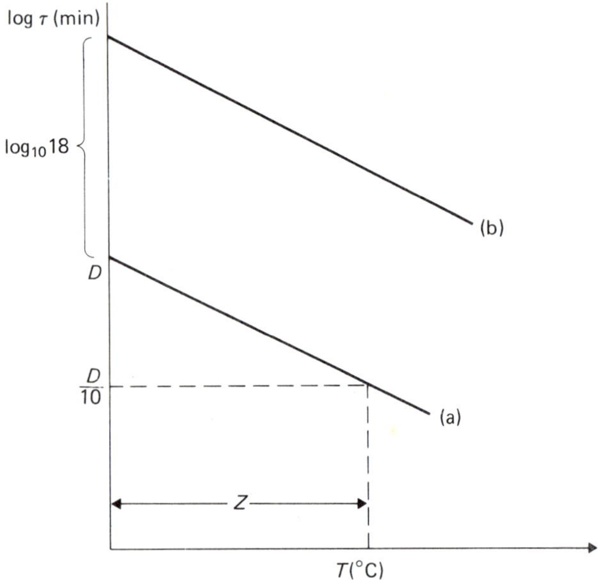

Figure 2.9   (a) The relation between decimal reduction time and temperature. (b) The relationship between τ, the time required to reduce the population by 10^{18} and temperature T

Table 2.14 Heat resistance of a variety of microorganisms

Microorganism	*Heating temp* (°C)	*D* (min)	*Z* (°C)	*Suspension medium*
Bacterial spores				
Clostridium thermo-saccharolyticum	132	4·4	6·7	Water
	132	3·3	10	Molasses
Bacillus stearothermophilus	115	22·6	7·1	Water
Bacillus megaterium	100	1	8·8	Water
Vegetative cells of bacteria				
Salmonella typhimurium	55·5	0·5	4·2	Ringer's solution
Actinomycete spores				
Thermoactinomyces vulgaris	100	11	—	0·001 M phosphate buffer pH 7·0
Yeast spores				
Saccharomyces cerevisiae	55	0·9	—	Phosphate buffer pH 7·0
Spores of filamentous fungi				
Aspergillus chevalieri	65	50	12·8	Sucrose/malt agar
Virus				
Adenovirus	55	17	2·5	Ice cream

When $Z = (T_2 - T_1)$ then $\log_{10}(D_1/D_2) = 1$ and $D_2 = \frac{1}{10}D_1$, thus Z represents the increase in temperature required to reduce the decimal reduction time D to one tenth of its value.

If the decimal reduction time D is plotted as a function of temperature, on semi-logarithmic coordinates, a straight line is obtained with a gradient $1/Z$ (Figure 2.9a).

The heat treatment needed for a batch of medium is dependent on the maximum probability of residual contamination which can be tolerated. Thus, for example, a bacterial spore surviving in a volume 1000 times greater than that of the fermenter used would give a probability of one batch in a thousand runs being contaminated. Yeast spores may be slightly more resistant than the vegetative cells of many bacteria but, in fact, neither these nor the spores of filamentous fungi and actinomycetes have very great heat resistance (Table 2.14).

Example of a calculation

A fermenter of $50\,\text{m}^3$ capacity is to be used with a medium containing corn steep liquor, a product which frequently has a high spore population. The crude corn steep liquor is found to contain 20×10^6 viable spores/ml; thus the entire fermenter will contain

$$20 \times 10^6 \times 50 \times 10^3 \times 10^3 = 10^{15} \text{ spores}$$

An acceptable risk of contamination is considered to be 99·9%, that is to say there will be a statistical expectation of only one contaminated run per 1000 fermentations. Thus the number of spores has to be reduced from 10^{15} to 10^{-3}. It will thus be necessary to heat for a period τ corresponding to $(15 + 3)D = 18D$. If τ (in minutes) is plotted against the temperature T on a semi-logarithmic scale the curve obtained is the same as that relating the decimal reduction time D against temperature but shifted up the ordinate by $\log_{10} 18$ (Figure 2.9b). This is because, if $\tau = 18D$, then

$$\log_{10} \tau = \log_{10} 18 + \log_{10} D$$

If F is the value of τ for $T = 121°C$, and Z is the increase in temperature required to reduce the period of heating by 90% then:

$$\frac{\log_{10} \tau_T - \log_{10} F}{121 - T} = \frac{\log_{10} \dfrac{\tau_T}{F}}{121 - T} = \frac{1}{Z} \tag{2.10}$$

This equation may be rewritten in two equivalent forms when Z and T are in °F:

$$\tau_T = F \times 10^{(250 - T)/Z} \quad \text{or} \quad F = \tau_T \times 10^{(T - 250)/Z} \tag{2.11}$$

A simple example of the construction of a sterilization chart

An acid medium containing thiamine is to be subjected to continuous sterilization in a plate heat exchanger. The treatment is required to assure the acceptable destruction of the spores of *Bacillus coagulans* the resistance characteristics of which have been previously established:

$$F = 0·7; \quad Z = 18$$

To simplify the calculation, it is considered that the time taken to raise the temperature to $T°F$, and to subsequently recool, does not significantly contribute to sterilization. There are tables giving the expression $10^{(T - 250)/Z}$ as a function of T and Z allowing one to calculate τ knowing F and Z:

Thus   at 245°F (118°C) $r = 80$ seconds

at 266°F (130°C) $r = 5·4$ seconds

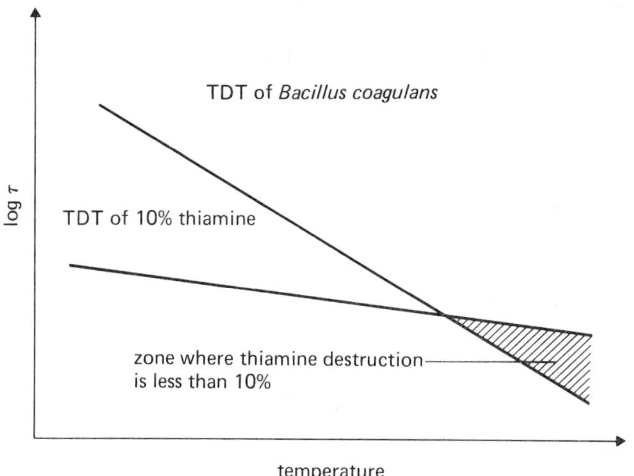

Figure 2.10   Determination of a zone of heat treatment reducing the numbers of *Bacillus coagulans* by 10^{18}, but destroying less than 10% of the thiamine in a culture medium

If it is required to preserve, for example, 90% of the thiamine in the medium then the curves corresponding to the destruction of *Bacillus coagulans* and to a rate of destruction of 10% thiamine are plotted on the same graph. The graph is then examined for the presence of a region fulfilling the two requirements—the desired level of destruction of spores and the destruction of less than 10% thiamine (Figure 2.10); this region is adopted as a time/temperature chart for satisfactory heat treatment, and the necessary holding time for a given temperature can be read off.

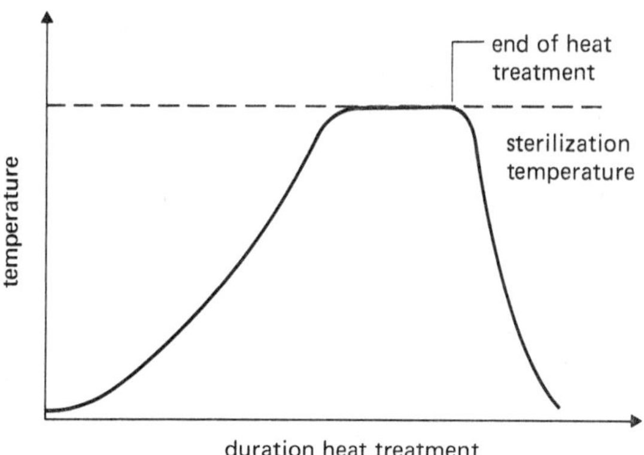

Figure 2.11   Change of temperature in a culture medium during the sterilization of a fermenter

Experiments of this kind demonstrate the virtual necessity for sterilizing at relatively high temperatures for very short times (i.e., continuous sterilization) if the medium contains heat labile constituents. When the heating and cooling phases of heat treatment are sufficiently long to contribute to sterilization then it is no longer possible to use calculations based on a constant temperature (Figure 2.11).

In a large industrial fermenter the duration of the periods during which the temperature is increasing and decreasing may have a more significant effect on the medium than the period at the highest temperature. For this reason the graphical method described below may be used.

Experimental determination of the heat resistance of microorganisms

It is required to determine the values of F and Z which will then allow the construction of a graph relating time of heat treatment and temperature, for F represents one point on that line and $1/Z$ represents the slope of the line (Figure 2.9). Although it may be desirable to determine the thermal resistance of unknown communities in the medium to be used, because this is where problems will arise in practice, it is usual to measure the resistance of a known organism in a defined medium. The thermophile *Bacillus stearothermophilus* FS 1518 is frequently taken as a reference organism because its spores are amongst the most resistant to heat treatment.

Preparation of spore suspension

The culture is grown on an agar medium favouring spore formation. After incubation, the sporing culture is suspended in a little sterile water and the suspension treated with lysozyme to destroy vegetative cells and the sporangia. The presence of spores is confirmed microscopically and the suspension transferred to a sterile flask containing glass beads. After shaking to break up any aggregates it is filtered through two layers of sterile cheese cloth. The number of viable spores is counted after treatment of a suitable dilution at 100 °C for 5 minutes, to destroy any surviving vegetative cells, and to activate germination of the spores.

Tube assay method

Tubes, closed at one end, and about 12 cm in length, are prepared from glass tubing of 1 mm thickness and 9 mm outside diameter. To each is added 2 ml of medium (using a syringe) and suspension of a known number of spores. The tubes are then sealed by flaming at about 5 cm above the surface of the contents, and then immersed in a water bath, or a bath of oil or ethylene glycol, stirred and maintained at the required temperature.

At known periods of time, tubes are removed, plunged into cold water, opened aseptically, and the contents used to inoculate a tube of broth in order to demonstrate survival or destruction (the result is scored $+$ or $-$). Plotting temperature along the abscissa and the logarithm of time along the ordinate two curves can be obtained which should be fairly close to one another (Figure 2.12).

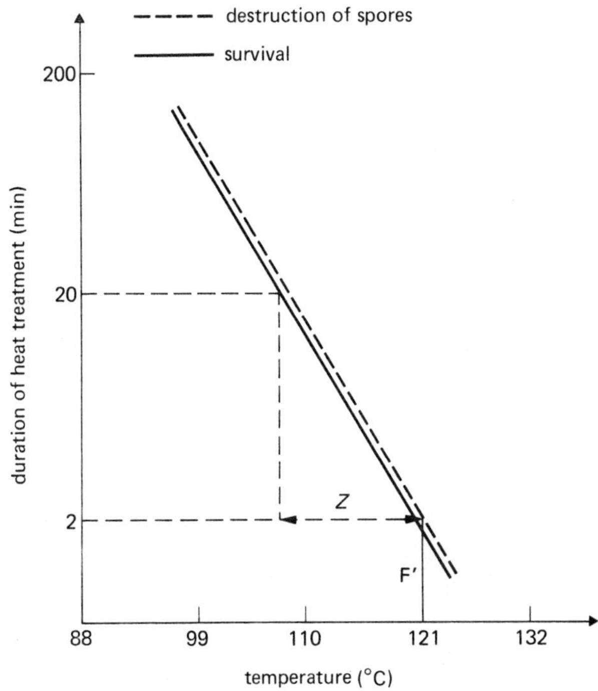

Figure 2.12   Curves for survival and destruction of spores

Example

The curve obtained gave the following results:

$$F' = 2 \text{ minutes}$$

$$Z = 15°C$$

This result is only relevant to the destruction of the number of spores used in the experiment (174/ml). The value of F corresponds to an 18-fold decimal reduction in the number of the same spores, therefore

$$F = \frac{18}{\log_{10} 174} \times 2 = 16 \text{ minutes}$$

This method is usable only if the experimental points obtained lie reasonably close to a straight line. Otherwise it is necessary, for each temperature, to obtain the proportion of surviving spores as a function of time from which the decimal reduction time D can be calculated and the value of Z deduced from a study of the relationship between D and temperature. It is necessary then to carry out a full viable count at a range of times at each temperature (Table 2.15).

Table 2.15 Survival of *Clostridium sporogenes* PA 3679 at 115°C (10,000 spores to tube)

Time of heating (min) (absolute number)	*Number of surviving spores per tube*	($\log_{10}$)
10	90	1·95
14	12	1·08
18	2	0·30
22	0·3	$\bar{1}$·52
28	0	—
36		—

Let a be the number of spores initially (e.g. 10,000) and b the number surviving after heating at 115°C for 10 minutes (e.g. 90).

$$D = \frac{10}{\log_{10} a - \log_{10} b} = \frac{10}{\log_{10} 10{,}000 - \log_{10} 90} = 4.88$$

By repeating the experiment at a number of temperatures the corresponding values of D can be calculated (Table 2.16). Finally, by plotting D (on a logarithmic scale) against temperature, Z can be determined from the following relationship.

$$\frac{\log_{10} D_1 - \log_{10} D_2}{(T_2 - T_1)} = \frac{1}{Z}$$

Knowing D it is easy to calculate F and with a knowledge of both F and Z for a given culture a working sterilization table can be established.

Table 2.16 Value of D as a function of temperature
for *Clostridium sporogenes* PA 3679

Temperature (°C)	D (min)
110	19·65
113	10·46
115	4·92
118	2·56
121	1·23

The classical technique of sterilizing using steam may be carried out in two ways:

1. *Batchwise in the fermenter*

In this method the medium is heated by direct injection of steam (it is necessary to allow for the volume of water condensed in formulating the final composition of the medium), or the container and its contents are indirectly heated by circulating steam through pipes immersed in the medium. Sometimes a fermenter may be constructed with a double wall and steam can be circulated between the two walls. These processes involve a considerable loss of production time for it may require several hours to reach a temperature of 120°C and subsequently cool the medium before inoculating; the alternative of using separate batch cookers for media sterilization, occupies increased plant space. Keeping a medium at these elevated temperatures for such long periods of time presents a number of problems, such as the denaturing of proteins, destruction of growth factors, polymerization of some constituents and darkening of the medium. The darkening usually arises from caramelization of sugars, oxidation of phenols and the Maillard reaction between the aldehyde group of reducing sugars and the amino group of amino acids giving products which are often toxic to microorganisms.

It is sometimes recommended that sugars and nitrogenous substrates be sterilized separately and mixed aseptically when required.

2. *Continuous sterilization*

This process gives advantages both in the saving of time and plant space and in improved quality of the medium. For example in the production of vitamin B_{12} an increase of 160% in the yield has been obtained. The medium is passed through a heat exchanger in which it is raised to the required high temperature in a very short time. It then passes to a holding coil where it is maintained at this temperature for a predetermined time and is finally rapidly cooled by

countercirculating in the exchanger against cool input medium, and then
against cold water. Because of the short holding times possible, sterilization
temperatures in excess of 120°C may be used without any detrimental effects

Table 2.17 Parameters involved in continuous sterlization

Details	Steam injection	Heat exchanger
Holding time	1–5 min	10 min
Holding temperature	140°C	120°C
Heating up/cooling down period	negligible	1% contribution

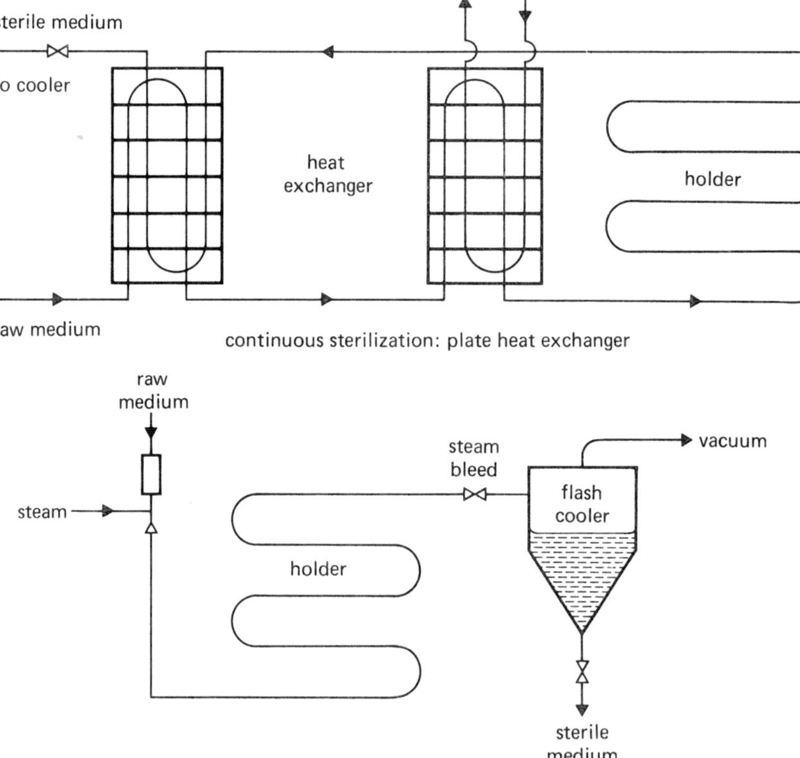

Figure 2.13 Schematic diagram for continuous medium sterilization

on the nutritional qualities of the culture medium (Table 2.17). A slightly different system frequently used involves the injection of high temperature steam into the medium, the required temperature ($< 140°C$) being reached in microseconds. After passing through a holding coil, the medium is flash cooled in a vacuum chamber (Figure 2.13).

It should be noted, in passing, that prolonged sterilization times can occasionally lead to increased yields, for example in the riboflavin fermentation.

2.4   Preparation of inoculum

It may be necessary to inoculate the final medium in such a way that the inoculum develops sufficiently rapidly to repress any possible contaminants. In order to obtain a large, vigorously growing inoculum, of the order of 5000–10,000 l for a fermenter some 10–20 times as large, it is necessary to prepare it in several stages. The laboratory will usually only have at its disposal a pure culture preserved frequently as a lyophilized sample.

A spore suspension is prepared, diluted and used to inoculate an agar slope. The increased number of spores obtained from the slope are used to inoculate a liquid medium in an Erlenmeyer or Fernbach flask which is incubated on a shaker; it is essential that the medium and conditions used give a large number of propagules by fragmentation of the mycelium. This is then passed on to inoculate a small fermenter (300–500 l) the growth from which is used to inoculate a larger fermenter (3000–5000 l) producing the working inoculum for the final production vessel. Using mobile stainless steel containers and sterile compressed air inocula may be transferred quickly and aseptically. All containers, tubes and valves are previously cleansed and then sterilized with steam.

At each stage rigorous control is necessary to avoid the three major causes of failure, contamination, bacteriophage and mutation. The presence of contaminants may be detected by frequent microscopic examination or by subculture. The presence of phages is usually apparent from the slowing down or arrest of growth. Control of mutations, which may be fairly frequent in the streptomycetes, is very difficult, and it may only be observed as a reduction in yield which may not always be very significant. As far as possible stable high yield mutants should be obtained and used as the starting culture. However it should be pointed out that overproduction of a particular metabolite, desirable though it may be from the fermentation chemist's point of view, is biologically disadvantageous to the organism. The latter will thus tend, by adapting to the artificial conditions of an industrial fermentation, to revert to a low-yield, rapidly multiplying type.

Bibliography

BORICK, P. M., (1968), 'Chemical sterilizers,' *Adv. Appl. Microbiol.*, **10**, 291–310.

CORRY, J. E. L., (1973), 'The water relations and heat resistance of microorganisms, *Progr. Ind. Microbiol.*, **12**, 73–108.

DEINDOERFER, F. H., WEST, J. M., (1960), 'Rheological properties of fermentation broth', *Adv. Appl. Microbiol.*, **2**, 265–273.

EINSELE, A., FIECHTER, A., (1971), 'Liquid and solid hydrocarbons', *Adv. Appl. Microbiol.*, **13**, 169–194.

FREDRICKSON, A. G., MEGEE, R. D., TSUCHIYA, H. M., (1970), 'Mathematical models for fermentation processes', *Adv. Appl. Microbiol.*, **13**, 419–464.

HUGO, W. B. (ed.), (1971), *Inhibition and Destruction of the Microbial Cell*, Academic Press, New York and London.

KEMPE, L. L., (1960), 'Sterilization of media for biochemical processes,' *Adv. Appl. Microbiol.*, **2**, 313–318.

KLUG, M. J., MARKOVETZ, A. J., (1971), 'Utilization of aliphatic hydrocarbons by microorganisms,' *Adv. Microbiol. Physiol.*, **5**, 1–39.

REESE, E. T., MANDELS, M., WEISS, A. H., (1972), 'Cellulose as a novel energy source,' *Adv. Biochem. Eng.*, **2**, 181–200.

RICHARDS, J. W., (1968), *Introduction to Industrial Sterilization*, Academic Press, New York and London.

SOLOMONS, G. L., (1971), 'Fermentation equipment', *Adv. Appl. Microbiol.*, **14**, 231–246.

3 The Formation and Extraction of Fermentation Products

This chapter concerns the control of the physico-chemical parameters of the culture during the course of the fermentation, and extraction of the required product when the process is complete.

3.1 Control of fermentation conditions

The large industrial batch fermenter is generally a tall cylindrical vessel with a height:diameter ratio of at least 5:2, constructed either of stainless steel or mild steel with a stainless steel or glass coating. Capacity varies according to process needs; a very large fermenter might have a capacity of some 300,000 l. Steam under pressure is the usual method of sterilization and fermenters are therefore subject to tests and standards similar to those used for boilers.

The liquid medium with which the fermenter is charged is usually at a low redox potential, having been heat sterilized. When a process requires anaerobic conditions, as in the production of ethanol using yeasts or (formerly) butanol and acetone using *Clostridium acetobutylicum*, any remaining traces of oxygen are rapidly taken up at the beginning of the fermentation, with the production of carbon dioxide and hydrogen.

The majority of industrial fermentations however employ aerobic organisms which require O_2 for rapid growth. Considerations of plant space usually preclude the use of shallow containers. Deep fermenters have to be used but here, mere diffusion of O_2 from the surface of the medium would be quite inadequate: some form of forced aeration is essential. In both aerobic and anaerobic fermentations the growth rate is also dependent on the rate of transfer of nutrients across the cell membrane, and some form of agitation of the culture is frequently applied to accelerate this.

Aeration and agitation

Because pure oxygen is expensive and requires specially designed plant, aeration is usually by direct injection of sterile air under pressure. Oxygen has a low solubility in aqueous media which decreases with increasing temperature and its availability to the microorganisms may thus be limiting. Indeed, if the concentration of oxygen in the medium falls below a critical value there may be metabolic disturbances leading to death of the microorganisms (Table 3.1).

Table 3.1 Maximum consumption and critical concentration of oxygen for various microorganisms

Microorganism	Temp (°C)	Critical conc (mmoles/litre)	Maximum consumption (mmoles/litre/hour)
Azotobacter	30	0·018	260
Escherichia coli	37	0·008	5–8
Yeast	30	0·004	10–15
Penicillium chrysogenum	24	0·022	20–30
Acetobacter	30	—	90
Streptomyces griseus	30	—	15

When cultures of *Acetobacter*, growing on ethanol for the production of acetic acid, are deprived of oxygen for as little as 10 seconds they lose their viability.

The presence or absence of oxygen often influences the nature of the metabolites produced. Thus *Serratia marcescens* produces its characteristic red pigment, prodigiosin, aerobically but, under anaerobic conditions, synthesizes the useful enzyme asparaginase. *Aspergillus niger* produces α-amylase under conditions of restricted O_2 supply, but for it to produce citric acid ample O_2 is needed. It should also be noted that the concentration of oxygen required for optimum growth may differ from that yielding optimum product synthesis. The oxygen requirement may be less for optimal product formation than for maximum growth, as in the case of the production of α-amylase by *Aspergillus niger*, or it may be more, as in the case of citric acid production by the same microorganism.

Transport of oxygen through the culture medium to the growing cells tends to be a critical factor. Table 3.2 gives specimen data for a yeast growing in a glucose medium.

Table 3.2 Consumption of oxygen by yeast (mineral salts + 1% glucose)

Substrate	Concentration in solution (ppm)	Critical concentration (ppm)	Rate of consumption (mmoles/g biomass/h)
Glucose	10,000	100	2·8
Oxygen	7	0·8	7·7

The oxygen requirement per minute for a yeast suspension containing 10 g/l is:

$$7{\cdot}7 \times 10 \times \frac{32}{60} = 41 \text{ mg/l}$$

This is well above the oxygen concentration which could be present in the medium and explains the necessity for active injection of air into the fermenter. In the transport of oxygen from the gaseous phase to the enzyme systems of the microorganism it is the transfer from the air bubble to the liquid phase which is rate limiting. In considering transport across the air/liquid interface it is possible to use Fick's law describing the diffusion of a solute across a membrane.

$$\frac{dN}{dt} = -Da\frac{dc}{dx} \qquad (3.1)$$

dN = number of molecules transferred during the interval of time dt (seconds).
D = diffusion coefficient ($cm^2\,s^{-1}$)
a = interfacial area (cm^2)
dc = change in concentration of solute (moles cm^{-3}) over a distance dx in the direction of flow across the membrane.

Since the last parameter is difficult to determine, coefficients $K_L = D/x$ and Δc (the total difference in concentration of the solute across the membrane of thickness x) are used. Equation (3.1) then takes the form:

$$\frac{dN}{dt} = -K_L a\,\Delta c \qquad (3.2)$$

Thus the rate of transfer of oxygen, expressed in mmoles/s, is given by an equation of the type:

$$(K_L a)(c^* - c_L) \qquad (3.3)$$

c^* = concentration of oxygen at the interface inside the bubble, corresponding to the theoretical value of a saturated solution for oxygen (at the temperature and partial pressure used)
c_L = concentration of oxygen in the medium surrounding the bubble of air.

For any particular fermenter assembly, $K_L a$ is a constant representing a volume transfer coefficient having the dimensions L^3T^{-1} and may be expressed, for example, in hours^{-1} (strictly speaking fermenter volume per hour). This coefficient can be experimentally determined by a method based on the fact that sodium sulphite reacts rapidly with oxygen in the presence of certain metal ion catalysts. The reaction is so fast, at least in comparison with the rate of transfer, that c_L may be taken as zero. Under these conditions the rate of transfer of oxygen approximates to K_Lac^* mmoles h^{-1}.

The fermenter is flushed with nitrogen to remove all air and is filled with a 0·5 M solution of sodium sulphite, containing traces of a copper or cobalt salt. It is then aerated at a given rate as it would be during a fermentation. Samples are removed at intervals into an atmosphere of nitrogen and the quantity of sulphite oxidized is estimated by allowing residual sulphite to react with excess iodine and back titrating with thiosulphate. By plotting the quantity of sulphite oxidized as a function of time a straight line is obtained, the gradient of which gives the value of the rate of oxygen transfer.

$$K_V = K_Lac^*$$

From a knowledge of c^*, the concentration in the medium saturated with oxygen, it is possible to calculate K_La.

In practice, values determined for the coefficient K_La should reflect the capacity for oxygenation of the medium in a particular fermenter but the results must be interpreted with caution: oxygen absorption by a suspension of microorganisms in a medium may be less than that suggested by the sulphite method. Values of the coefficient K_La for some different growth vessels are given in Table 3.3.

Table 3.3 Volume transfer coefficient K_La for different types of vessel

| | Stirring | | |
| | Amplitude | | K_La |
Type of vessel	(cm)	rev/min	(h^{-1})
25 ml Warburg flask with 3 ml medium	2·4	150	8·3
1 l Erlenmeyer flask with 300 ml medium (reciprocal shaker)	7·0	9	24
250 ml Erlenmeyer flask with 75 ml medium (rotary shaker)	3·75	220	26
500 ml Erlenmeyer flask with 50 ml medium (rotary shaker)	2·9	253	200
Fermenter (15 cm diameter) with 7·5 cm diameter stirrer	—	500	420
Fermenter (58,000 l) 36 CV/h power input	—	—	370

Another method for measuring oxygen transfer employs membrane electrodes which allow the continuous measurement of dissolved oxygen. Physical factors such as the temperature and viscosity of the medium also affect oxygen transfer; their influence can be determined experimentally or calculated by somewhat complex formulae. Moreover, the bubbles of air change in composition as they pass through the medium; allowance should be made for this by calculating a mean value of oxygen transfer.

Aeration is inseparable from agitation, which accelerates oxygen transfer by breaking up air bubbles, increasing their relative surface area, and by introducing turbulence, which delays the rising of the bubbles. Design of the aeration/agitation system depends on the process. When the culture is of low viscosity, the introduction of air through a sparger at the bottom of the fermenter may well provide both adequate aeration and agitation. (The size of the perforations in the air sparger is important for if they are too small blockage may occur and a high pumping pressure is required whereas if they are too large oxygenation may be insufficient.) If the medium is viscous, or thickened by mycelial growth, oxygen transfer is impeded: the air introduced at the bottom of the fermenter has to be broken up by a propeller–stirrer and its rise to the surface, already slowed by turbulence, further reduced by baffles fitted to the fermenter wall.

The quantity of air needed is usually large, often of the order of the volume of the fermenter per minute. The bubbles of air occupy a considerable proportion of the volume of the medium compared with that taken up by the microbial cells, yet, because of their larger size, their surface area is proportionately low (Table 3.4).

Table 3.4 Comparison of the relative surface area/volume ratio of air bubbles and various microorganisms in suspension in an aerated and agitated culture medium

Microorganism	Dimension (μm)	Relative ratio (area/volume)	Number of particles/ml	% of volume of medium
Bacteria coccus	1	1000	10^9	0·05
rod	5 × 1	710	19^9	0·3
Yeast (spherical)	5	200	10^8	0·6
Filamentous actinomycete	2	330	—	0·2
Filamentous mould	10	66	—	2·0
Air bubble	1000	1	380	19

Sterilization of air

Air contains microorganisms, mostly bacteria or fungal spores, which may be suspended free or borne on dust particles. In polluted urban areas, where fermentation plant is often situated, it is possible to find an average of 10^3–10^4 viable organisms per cubic metre of air.

Both heat treatment and filtration can be used to sterilize air. Complete sterilization by heat alone would require passing air through heat exchangers at 260–300°C; supplementary cooling would also be needed. The two methods can be combined: air is first compressed, the heat produced causing the destruction of at least part of the contaminating flora, and then partly decompressed before being passed through filters (adiabatic compression).

However it is more usual to employ low-pressure filtration alone. Fibre glass filters, or fibre glass impregnated with resin, are widely used. They function mainly by electrostatic charge effects and are much more effective than cotton filters which were formerly used. Moreover, they can be cleansed and re-used, are less prone to packing effects during use, and are resistant to microbial attack.

To evaluate performance, it is assumed that each particle which touches a glass fibre remains attached to it and that there are no preferred pathways through the filter. It is then possible to assume a uniform concentration of particles at any given depth through the filter and the diameter of the filter is not involved in the calculations.

Under these conditions a unit thickness of a filter column will reduce the concentration of particles by a constant proportion. If N is the concentration of particles in the air at a depth x through the filter column, and N_0 is the original concentration,

$$\frac{\mathrm{d}N}{\mathrm{d}x} = -KN \tag{3.4}$$

$$\ln\frac{N}{N_0} = -Kx \tag{3.5}$$

$$\frac{N}{N_0} = e^{-Kx} \tag{3.6}$$

The efficiency of the filter, n'_0, which represents the ratio of the number of particles removed to the total originally present, is given by:

$$n'_0 = \frac{N_0 - N}{N_0} = 1 - e^{-Kx} \tag{3.7}$$

For comparing different filtration materials the thickness X_{90} which will remove 90% of the total number of particles, is calculated from equation (3.5).

$$\log_{10}\frac{1}{10} = -\frac{K}{2\cdot3}X_{90}$$

$$\therefore \quad X_{90} = \frac{2\cdot3}{K}$$

Values of X_{90}, determined for different microorganisms, a range of filter materials, and as a function of air flow, are shown in Table 3.5.

Table 3.5 Values for the thickness of filter materials which will retain 90% of particles in air

Filter material	Diameter of fibres (μm)	Microorganism	Air speed (cm/s)	X_{90} (cm)
Glass wool	16	Spores of *Bacillus subtilis*	3	4
			15	9
			30	11·5
			150	1·5
			300	0·4
	18·5	*Serratia marcescens*	3·9	3·1
			78	3
Glass fibre	8·5	*Escherichia coli* (phage tyle T2$_2$)	3	0·4
			15	0·6
			30	0·7
			150	0·8
			300	1·1
Norite (15–30 mesh)	—	Spores of *Bacillus cereus*	1·4	1·7
			6·4	1·5
Activated carbon (4–8 mesh)	—	Spores of *Bacillus cereus*	18	1·7
			28·5	8·7

Table 3.5 shows that it is usually advantageous to use high flow rates indicating the importance of impaction as a mechanism for removal of bacteria. It is important that the air be dry, high humidity causing condensation in the filter which may support microbial growth.

Because aerial contaminants may rapidly overgrow the fermentation medium, it is practically essential that the conditions of filtration should not allow a single propagule to pass throughout the process and this may require an efficiency of at least 99·99% (Table 3.6).

For example, it might be required to find what thickness of glass wool was necessary to protect a 75,000 l fermenter for 90 h during aeration with 22 m³/min air containing 200 spores/m³, at a flow rate of 0·3 m/s. Using formula (3.5) and the value of X_{90} given in Table 3.5:

$$\log_{10}\frac{N_0}{N} = \frac{K}{2\cdot3}x = \frac{x}{X_{90}}$$

Table 3.6   Values for the time taken for air contaminants to
invade 5 l and 75,000 l fermenters

Growth stage and growth rate μ of contaminant		Time (h) *to reach a level of contamination of*:			
		$(10^6/\text{ml})$		$(10^8/\text{ml})$	
		5 l	75,000 l	5 l	75,000 l
$\mu = 2\,\text{h}^{-1}$	Exponential phase	16	23	19	26
	After a stationary phase of 6 h	22	29	25	32
$\mu = 0{\cdot}05\,\text{h}^{-1}$	Exponential phase	65	92	78	106
	After a stationary phase of 6 h	71	98	84	112

$$N_0 = 22 \times 60 \times 90 \times 200 = 2{\cdot}4 \times 10^7$$
$$N = 10^{-3} \text{ (corresponding to the probability of 1 in 1000 of a spore crossing the filter)}$$

thus

$$x = 11{\cdot}5 \times \log_{10}(2{\cdot}4 \times 10^{12}) \sim 120\,\text{cm}$$

Foam control

The intense aeration and agitation in a fermenter may lead to abundant foaming of the medium which may overflow through air egress and sampling ports, increasing the possibility of contamination. Foaming often occurs at a characteristic stage in a fermentation: early foaming is due to medium constituents, later foaming is usually due to microbial products. Measured doses of antifoam agents, such as vegetable or animal oils, long chain alcohols, silicones and selected detergents, may be added automatically in response to a signal from an electronic foam sensor which responds to the rising foam level in the fermenter. An excess of antifoam agent is undesirable, particularly when foam flotation is later to be used for separating cells from the spent medium. Sometimes the antifoam agent acts as an additional nutrient augmenting the yield of the required product. Such is the case when soya oil is used during the production of riboflavin; in early penicillin processes cottonseed oil augmented yields.

pH control

Each microbial species has an optimum pH for growth; moreover, microbial metabolism usually causes a change in pH of the medium. Some form of pH control is thus essential. Metabolism of sugars usually results in the production of organic acids; however, utilization of potassium nitrate and many other nitrogen sources tends to make the medium become alkaline; in the balance, most fermentations lead to progressive fall in pH unless this is controlled.

The production of any particular metabolite is equally likely to require a narrow range of pH. For example, *Clostridium acetobutylicum* produces acetone and butanol only in an acidic medium. Production of penicillin G by *Penicillium chrysogenum* is optimal only when the pH is carefully controlled (see Figure 3.1).

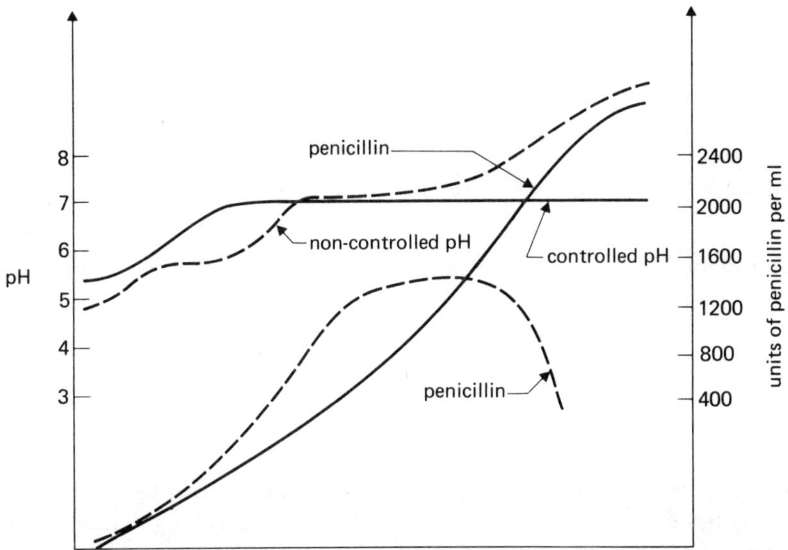

Figure 3.1   Production of penicillin with and without pH control

The cost of buffers (such as phosphates) to control pH is negligible for laboratory work but would be near-prohibitive industrially. (Calcium carbonate is sometimes used, e.g. in lactic acid production.) Cheaper materials, hydrochloric acid, sodium hydroxide or ammonia, are usually used to adjust the pH, being added according to the results of successive titrations or in response to a signal from a pH electrode immersed in the medium.

Temperature

All microorganisms display an optimum growth temperature, which is usually close to the maximum temperature allowing growth (Figure 3.2).

Many bacteria and fungi used industrially have an optimum temperature of about 30°C. However, thermophilic strains, requiring 45–65°C, are of

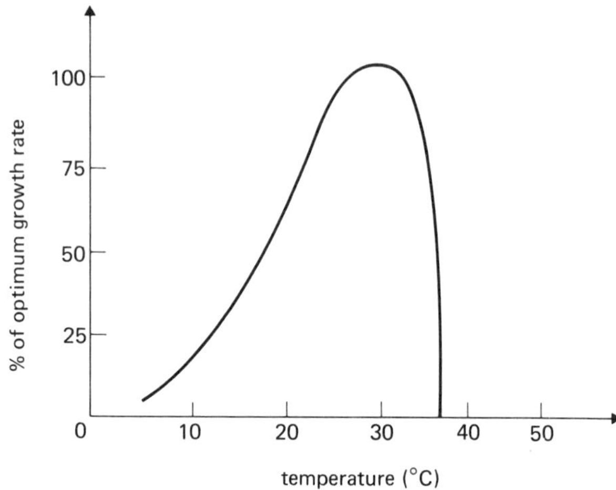

Figure 3.2   Rate of growth of a culture of a mesophilic bacterium as a function of
temperature

considerable interest because such temperatures considerably reduce the risk
of contamination. Sometimes a process may be programmed through two
different temperature ranges to obtain optimum yield: in the case of penicillin
production, the initial phase during which mycelium is produced requires a
lower temperature than the subsequent phase during which the antibiotic is
synthesized.

From a thermodynamic point of view, a growing microorganism is a system
in which energy produced by catabolic reactions is used to drive a series of
anabolic reactions. However, only a part of the total energy available from the
substrate is theoretically available to drive anabolic processes.

Thus, the energy used for growth represents only a fraction of the energy
released (29% in the case of an alcoholic fermentation) the rest being released
in the form of heat. The amount of heat released depends partly on the
substrate utilized (Table 3.7).

Table 3.7   Heat released during the continuous culture of microorganisms using
different substrates (the dilution rate is $0.10\,h^{-1}$ for a cell density of $15\,g/l$)

Substrate	Cell yield as % of substrate used	Heat released (kcal 100g cell)	(kJ/100 g cell)	Rate of heat released (kcal/h)	(kJ/h)
sucrose	50	380	1590	6	25
n-alkanes					
$(C_{12}-C_{18})$	100	780	3270	12	50
Methane	60	1800	7550	30	125

In order to maintain a constant temperature during the process it is thus usually necessary to provide some heat at the start of the fermentation and an efficient cooling system later in the process. A cooling coil, or a double walled fermenter, allowing circulation of water, will permit temperature control to within 0·2°C.

3.2   Principal types of culture

Industrial fermentations may be carried out in numerous ways:

Solid or semi-solid media

These are generally avoided in Western countries because they occupy too much plant space and may create harvesting difficulties, but they are still widely used in Japan. Those used include such materials as brans soaked in nutrient solutions. In the production of fungal amylases from *Aspergillus oryzae*, a mixture of bran and starch is moistened with a mineral solution containing some hydrochloric acid and thinly spread out in trays which are autoclaved (some starch is hydrolyzed) cooled and inoculated with a spore suspension. After several days at 30°C a sporing mass of mycelium develops. The whole may then be dried and ground to give a crude preparation of amylase at relatively little cost; alternatively a semi-purified amylase can be prepared by extraction and precipitation with ammonium sulphate or ethanol.

Batch liquid culture

The inoculated culture is grown in a batch fermenter with aeration and stirring; the fermenter is then emptied and the required product extracted. Such a process is characterized by a number of rate reactions (Figure 3.3) represented by the following formulae:

$$\text{rate of growth } \mu = \frac{1}{X}\frac{\mathrm{d}X}{\mathrm{d}t} \tag{3.8}$$

$$\text{rate of production } K' = \frac{1}{X}\frac{\mathrm{d}P}{\mathrm{d}t} \tag{3.9}$$

$$\text{rate of substrate utilization } K'' = \frac{1}{X}\left(-\frac{\mathrm{d}S}{\mathrm{d}T}\right) \tag{3.10}$$

X = cell concentration
P = product concentration
S = substrate concentration

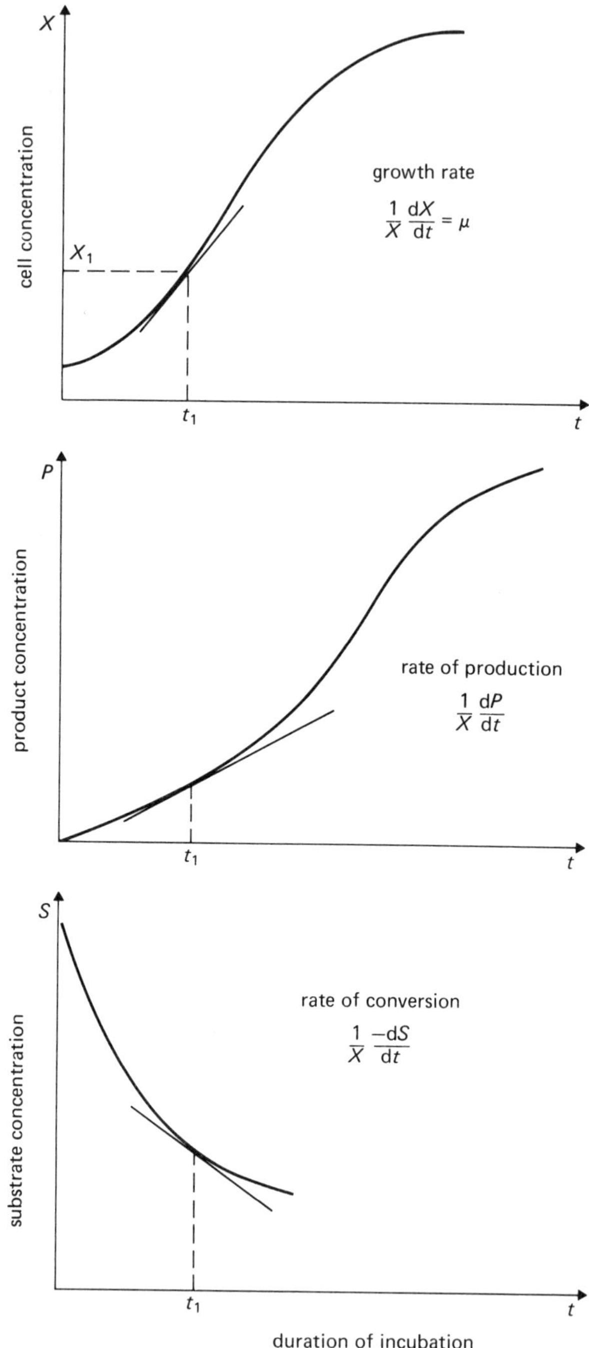

Figure 3.3   Production of biomass X, of product P and the consumption of substrate S all as a function of time

In the case of bacteria or yeasts each cell yields two identical cells following asexual reproduction; after n generations the initial number X_0 at time $t_0 = 0$, becomes X where:

$$X = X_0 2^n \tag{3.11}$$

If n is the number of divisions in time t the generation time g is given by t/n

$$\therefore \quad n = \frac{t}{g}$$

and
$$X = X_0 2^{\,t/g} \tag{3.12}$$

$$\ln X = \ln X_0 + \frac{\ln 2}{g} t \tag{3.13}$$

The term $(\ln 2)/g$ is known as the exponential growth constant μ:

$$\mu = \frac{0 \cdot 69}{g}$$

and equation (3.11) can be written in the form:

$$X = X_0 e^{\mu t} \tag{3.14}$$

and
$$\frac{\mathrm{d}X}{\mathrm{d}t} = \mu X \tag{3.15}$$

During the exponential phase, the growth rate reaches its maximum value $\mu_{\max}$. If the concentration of a single substrate has a limiting effect, the growth rate will be less than $\mu_{\max}$ and, as originally shown by Monod, it is possible to relate the growth rate and the concentration of the rate-limiting substrate by the following formula:

$$\mu = \mu_{\max} \frac{[S]}{K_s + [S]} \tag{3.16}$$

μ = growth rate constant
$[S]$ = concentration of limiting substrate
K_s = value of substrate concentration at which the growth rate is half the maximum growth rate $\mu_{\max}$.

This relationship is shown graphically in Figure 3.4. The Lineweaver–Burk plot (Figure 3.5) has the advantage of allowing determination of K_s by a simple graphical measurement.

With many bacteria, the value of K_s for glucose (as both carbon and energy source) is low, of the order of 10–100 mg/l. For an amino acid K_s may be even lower (c. 1 μg/l). In the case of the yeast *Torulopsis* K_s for xylose is of the order of 0·3 g/l.

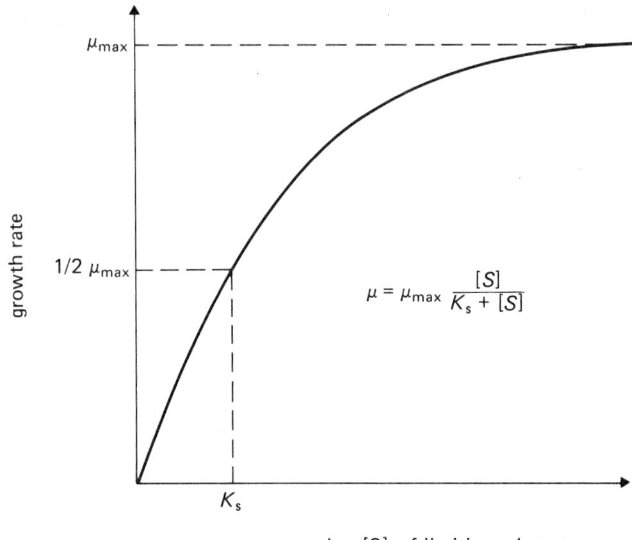

Figure 3.4   Growth rate μ as a function of concentration of the limiting substrate

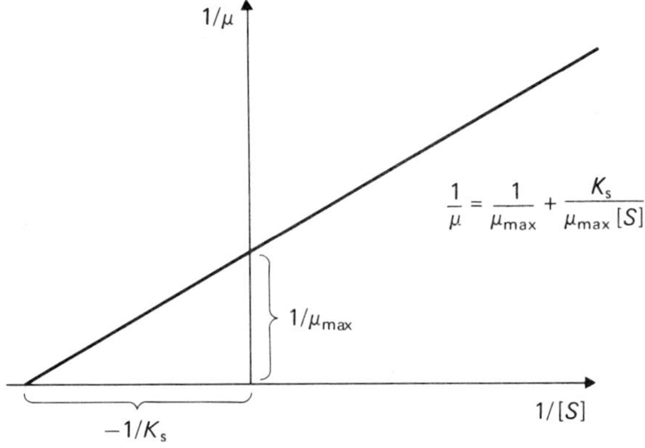

Figure 3.5   Lineweaver-Burk plot of growth rate μ as a function of the concentration [S] of rate–limiting substrate

Continuous culture

During the exponential phase of microbial growth, the medium becomes depleted of nutrients whilst products of microbial metabolism accumulate. If the medium in the fermenter is displaced by the addition of fresh medium and removal of spent medium and microbial cells, a culture may be maintained indefinitely in a steady exponential phase. This is what is meant by continuous culture (Figure 3.6).

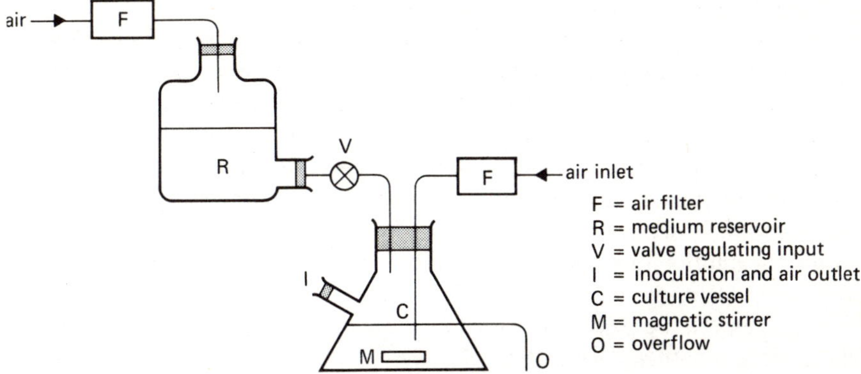

Figure 3.6   Schematic representation of a chemostat

The theory of continuous culture was established by Monod and by Novick and Szilard in 1950. If V is the culture volume and F the rate of addition of fresh medium the dilution constant D is given by:

$$D = \frac{F}{V}$$

In a continuous culture system two opposing phenomena affect the number of the population X:

(i) The division of the cells increases X at a rate dependent on μ_{max}

$$\frac{dX}{dt} = \mu X \tag{3.17}$$

(ii) Addition of fresh medium decreases X at a rate dependent on the dilution rate D

$$\frac{dX}{dt} = -DX \tag{3.18}$$

The resulting change in the population will be the sum of these two processes:

$$\frac{dX}{dt} = (\mu_{max} - D)X \tag{3.19}$$

There are three possible situations to consider:

(i) The dilution rate is greater than the maximum rate of division of microorganisms $D > \mu_{max}$. Under these conditions the microorganisms are unable to compensate by division for the dilution of the culture by fresh medium $dX/dt < 0$ and the population declines to zero.

(ii) The dilution rate is adjusted in such a way that $D = \mu_{max}$. Thus $dX/dt = 0$ and the microorganisms continue to grow at their maximum rate. This situation is realized where the fermenter operates as a *turbidostat*; the rate of dilution is regulated by the microbial density, monitored optically using the signal from a photoelectric cell to control the flow of fresh medium. Thus, if the amount of biomass exceeds a predetermined value, the flow is increased; if it falls, the flow is decreased.

(iii) The dilution rate is less than the growth rate, $D < \mu_{max}$; here, $dX/dt > 0$ and the population increases. However, the concentration of some rate limiting factor in the fermenter decreases preventing growth at maximum rate. The growth rate then decreases until it reaches a value $\mu = D$ at which point $dx/dt = 0$. Such a situation is achieved when the fermenter functions as a *chemostat*: the population is self-regulating at a level determined by the concentration of the limiting factor in the culture vessel.

It may be required to know the concentration S of the limiting substrate in the culture vessel during equilibrium. D is the dilution rate constant, S_0 the concentration of substrate in the fresh medium, S the concentration of substrate in the effluent (presumed to be equal to that in the culture vessel), and X the equilibrium population of cells in the culture vessel.

Since all substrate entering the culture vessel must be accounted for, the yield Y for the organism is given by

$$Y = \frac{X}{(S_0 - S)}$$

$$X = Y(S_0 - S) \tag{3.20}$$

$$dX = -Y\,dS \tag{3.21}$$

The rate of change dS/dt of the substrate concentration in the culture vessel may be expressed as:

$$\frac{dS}{dt} = \frac{dS}{dX}\frac{dX}{dt} \tag{3.22}$$

but
$$\frac{dX}{dt} = (\mu - D)X \tag{3.19}$$

and
$$\frac{dS}{dX} = -\frac{1}{Y} \tag{3.21}$$

$$\therefore \quad \frac{\mathrm{d}S}{\mathrm{d}X}\frac{\mathrm{d}X}{\mathrm{d}t} = -\frac{1}{Y}X(\mu - D) \tag{3.23}$$

or

$$\frac{\mathrm{d}S}{\mathrm{d}t} = \frac{X}{Y}(D - \mu) \tag{3.24}$$

As Monod has shown that: $\qquad \mu = \mu_{\max}\left(\dfrac{S}{K_s + S}\right)$

it is possible to write:

$$\frac{\mathrm{d}S}{\mathrm{d}t} = \frac{X}{Y}\left[D - \mu_{\max}\left(\frac{S}{K_s + S}\right)\right] \tag{3.25}$$

At equilibrium $\mathrm{d}S/\mathrm{d}t = 0$

$$D = \mu_{\max}\left(\frac{S}{K_s + S}\right)$$

$$D(K_s + S) = \mu_{\max}S$$

$$DK_s = (\mu_{\max} - D)S$$

or

$$S = K_s\frac{D}{\mu_{\max} - D}$$

This equation indicates that the equilibrium concentration of limiting substrate depends only on the dilution rate and not on the concentration S_0 in the fresh medium. If K_s is low, indicating a high affinity for the substrate, the value of S is also low so that the cell numbers at equilibrium is approximately

$$X = YS_0$$

These formulae make it possible to construct curves showing the equilibrium substrate concentration and cell numbers, and the rate of biomass production $P = DX$, all as functions of the dilution rate D (Figure 3.7).

Continuous cultures in a chemostat are of particular interest because, once equilibrium has been achieved, the rate of growth and the rate of substrate utilization are constant. Also the ratio A of the yield per unit time from continuous and batch cultures is high (Table 3.8).

Continuous culture thus yields, on average, about five times that obtained per unit time from a batch culture, and the extra yield is most marked with organisms having a high growth rate constant.

In practice, continuous culture techniques present special problems in avoiding contamination. Rigorously aseptic techniques have to be maintained throughout (in particular the sterilization of air and ensuring that no contamination *via* joints, valves and stirrer is possible). In the case of

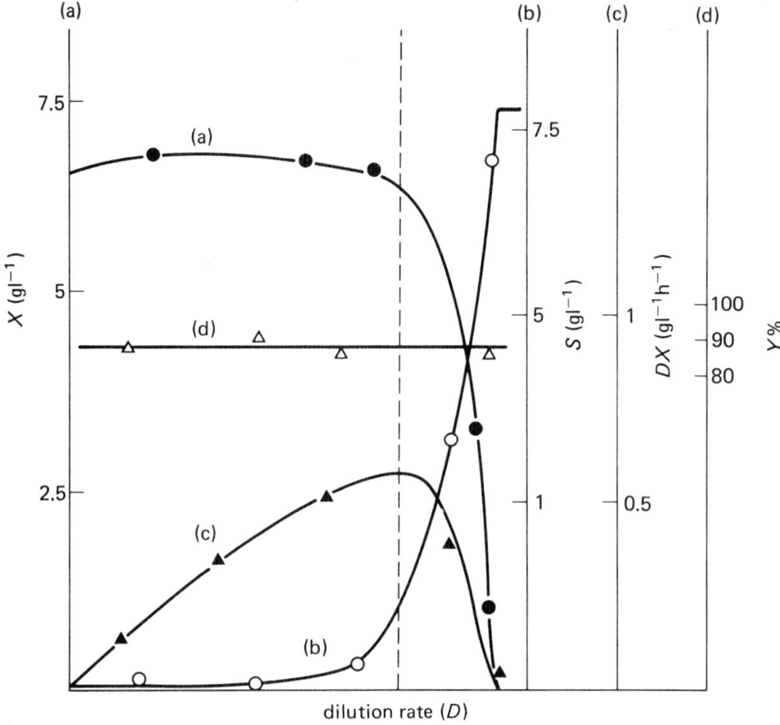

Figure 3.7   Various parameters as functions of the dilution rate D for a continuous culture of a yeast in a mineral medium (pH 4.0) containing 1% of a mixture of n-alkanes $(C_{10}-C_{18})$

Table 3.8   Variation of the yield ratio A of the yield per unit time for different values of the growth rate (using 5% inoculum and 6 h growth period for each batch culture)

Growth rate constant μ	Generation time in hours	A = $\dfrac{\text{continuous production}}{\text{batch production}}$
1·38	0·5	11·3
0·69	1·0	7·1
0·34	2·0	5·2
0·13	3·0	4·4
0·07	4·0	4·0
0·04	5·0	3·8

streptomycetes and filamentous fungi there is some additional difficulty in arranging the continuous removal of a spent medium containing stranded mycelia.

A major difficulty may arise because of the tendency of continuous culture processes to select those mutants having an increased growth rate.

When biomass or primary products related to growth (e.g. lactic acid) are required, a single stage simple chemostat is employed. For the production of secondary metabolites it may be better to use a multi-stage system of several fermenters arranged in series. The fresh medium is fed into the first chemostat, allowing growth of the organism: the effluent passes into a second stage where accumulation of product occurs. The derivation of equations to describe such systems is very complicated.

At the present time continuous culture processes tend to be most used industrially in situations where low pH of the medium discourages growth of other than the required organism, for example the production of yeast, beer or lactic acid, or where the growth substrate is suitable for the growth of the specific organism only, as in the production of acetic acid from ethanol, and of single cell protein from methanol.

Culture of microorganisms by dialysis

In this technique, the apparatus consists essentially of two compartments separated by a membrane: in one the microorganisms are maintained in suspension in a liquid medium, in the other, generally much larger, fresh

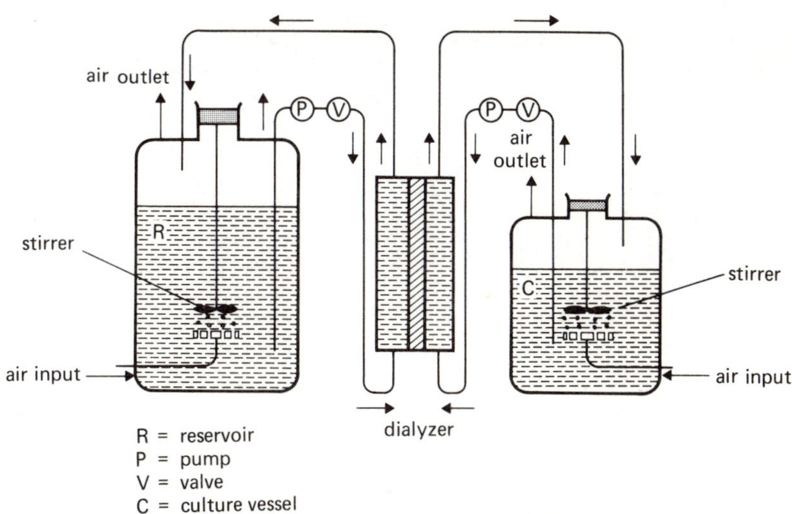

Figure 3.8   Schematic representation of dialysis culture

medium is stored. Nutrients pass across the membrane from the reservoir to the culture compartment and certain metabolites pass from the latter to the former (Figure 3.8).

Regulation of the system depends on the properties of the membrane and the ratio of the volumes of the two compartments. Dialysis systems may be used as a batch process or as a form of continuous culture. Usually the two compartments are standard fermenter vessels separated by a dialyzing unit through which the culture and medium are continuously circulated, separated only by the membrane. Types of membrane fall into three categories:

(i) Dialyzing membranes proper, composed of collodion, cellophane or a variety of plastics. Their pore size varies from 0·3–10 nanometres (1 nm $= 10^{-9}$ m) which completely retains large molecules such as enzymes and toxins, allowing the passage of small molecules such as those of sugars and mineral salts.

(ii) Membranes of cellulose acetate or plastic, porcelain or sintered glass which, strictly, are filter membranes. Their porosity varies from 25–200 nm and they thus only retain microbial cells allowing the passage of macromolecules.

(iii) Solution transport membranes made of teflon which allow the passage of gas molecules only. They do not have a porosity in the physical sense but act as a solvent in which soluble molecules pass by simple diffusion.

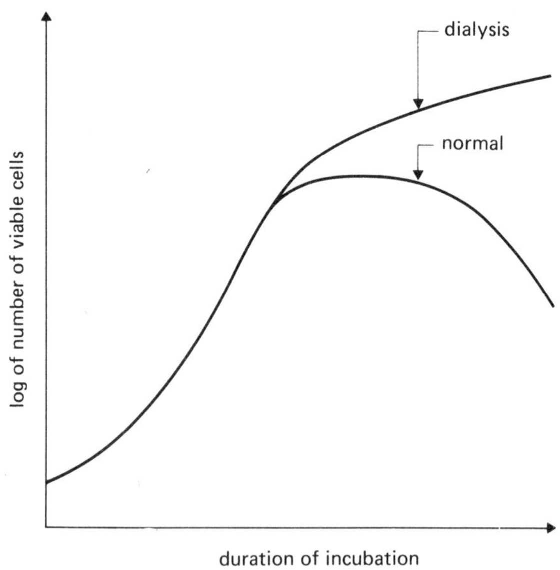

Figure 3.9   Growth curve of a normal batch culture compared with that for a dialysis culture

Dialysis cultures have two special characteristics which can be demonstrated experimentally. Firstly, there appears to be an extension of the exponential phase (Figure 3.9), probably due to the fact that a continued supply of nutrient materials is provided whilst potentially toxic metabolites are removed. It may also be noted that cells remain viable for a very long time during the stationary phase.

This first characteristic may be useful when the required product is formed in largest quantity during the exponential phase, for example, enzymes involved in primary metabolism. The other interesting characteristic of dialysis culture is the considerably increased concentration of biomass which can be achieved. The supply of nutrients should be a function of the ratio (in Figure 3.8 R/C) of the volumes of the two compartments which may be as high as 10. Excellent gas exchange is also possible (1 mmole O_2/l/min). Thus in the case of *Serratia marcescens* more than $2 \cdot 5 \times 10^{12}$ cells/ml could be obtained after 48 h incubation at 30°C. The high yield of concentrated biomass is of particular value in the production of vaccines; moreover, where type 1 membranes are used, extracellular protective antigens, such as toxins, are also retained, whilst media components remain in low concentration.

Some laboratory and industrial applications of dialysis culture techniques

The elimination of toxic metabolites can be very advantageous in certain kinds of metabolic study. For example, *Thiobacillus thiooxidans* behaves as an obligate autotroph in that, though able to metabolize glucose, it will not grow in mineral salts/glucose media. In fact it produces pyruvate which is toxic at a level of $0 \cdot 2$ mmole/l. If this organism is cultured by dialysis, the pyruvate produced passes into the reservoir compartment and the organism will grow in the presence of glucose at a faster rate than it does under strictly autotrophic conditions.

Lactobacillus delbrueckii produces lactic acid from glucose but the acid inhibits growth at concentrations greater than 10 g/l. In dialysis culture, the concentration of lactic acid can be maintained below the inhibitory level and the conversion of glucose to lactic acid considerably increased. In biosynthesis of threonine by an auxotrophic mutant of *Escherichia coli*, continuous removal of threonine by dialysis culture prevents feedback inhibition of the enzyme homoserine kinase (Figure 3.10) thus allowing an increased yield.

In the production of salicylic acid by naphthalene cleavage (Figure 3.11) using a strain of *Pseudomonas fluorescens*, the yield in a normal fermenter is limited by the toxicity of the product to 10.5 g/l, usually achieved after 48 h at 30°C. By comparison, after 15 days in dialysis culture, up to 200 g/l can be obtained. Despite the lengthy incubation period, this is still $2 \cdot 6$ times more

Industrial Applications of Microbiology

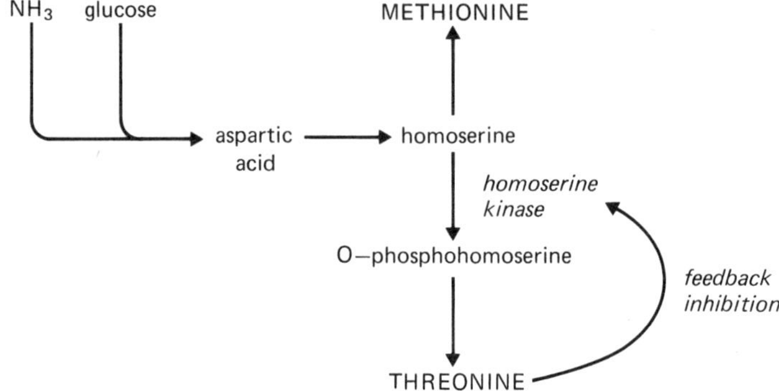

Figure 3.10   Production of threonine by a mutant of *Escherichia coli*

than could be achieved by setting up batch cultures every other day for the same period of time.

The possibility of obtaining cells completely free of high molecular weight, antigenic media constituents is useful in antigenic structure studies and would have practical application in vaccine production. Thus by incubating a dialysis bag containing an inoculum of *Brucella abortus* in a suitable nutrient solution, a thick suspension of very clean, viable cells containing no antigens other than those produced by the bacterium, can be obtained.

Table 3.9   Non-diffusible products obtained by dialysis culture

Product	Bacteria or cells
Hyaluronidase	*Streptococcus pyogenes*
Levans	*Bacillus subtilis*
Anthrax toxin	*Bacillus anthracis*
Botulinum toxin	*Clostridium botulinum*
Diphtheria toxin	*Corynebacterium diphtheriae*
Enterotoxin	*Staphylococcus aureus*
Gangrene toxin	*Clostridium chauvoei*
Haemolysin	*Staphylococcus aureus*
Tetanus toxin	*Clostridium tetani*
Adenovirus	KB cells
BAI virus	Avian myeloblasts
Poliovirus	Hela cells

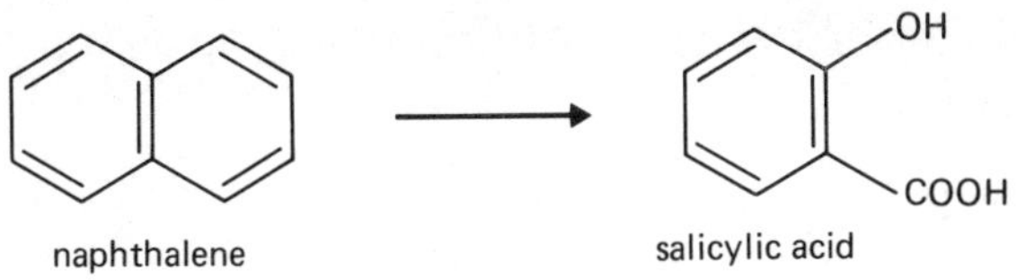

Figure 3.11

Dialysis culture also facilitates the recovery of macromolecules, such as enzymes, endo- or exotoxins or even viruses (Table 3.9). Thus it is possible by this technique to obtain eighty times more botulinum toxin than with the conventional methods. It is also possible to obtain fairly concentrated cell-free preparations of extracellular macromolecular products by introducing an intermediate chamber, separated from the nutrient reservoir by a dialyzing membrane and from the culture reservoir by a simple filter membrane. In this way seven times more staphylococcal enterotoxin and six times more streptococcal hyaluronidase have been obtained than by conventional methods: two types of hyaluronidase are produced, but in different proportions, suggesting that dialysis culture may modify the metabolism of the microorganism as well as allowing increased growth.

High cell yields also occur when mammalian or other vertebrate cells are cultured using dialysis techniques with appropriate media, sometimes of the order of 2×10^8 cells/ml. High density animal cell cultures can be used for high yield propagation of suitable viruses (e.g. poliovirus, adenovirus).

The use of gas permeable membranes can facilitate the culture of fragile microorganisms, (e.g. protozoa) which are sensitive to direct air injection or to antifoam agents. Another use is in the production of vaccines, where direct injection of air under pressure may generate safety hazards because of the formation of aerosols rich in infective pathogens; the use of gas permeable membranes obviates this risk.

The use of dialysis culture techniques allows the rational study of the effect of medium constituents on the behaviour of microorganisms. During anaerobic digestion of sewage, two microbial processes occur simultaneously, the first stage leading to the production of volatile organic acids, the second to the formation of methane. Experimentally it is possible to separate the two stages in a chamber divided by a polyvinyl dialyzing membrane, by maintaining the temperature and pH of each compartment at the optimum for the particular stage. By this method it has been shown that the temperature and pH optima are different for the two stages, that rate of acid production is the limiting factor for methane yield, and that the yield can be optimized by control of temperature, pH and redox potential; the bacteria responsible for each stage have also been isolated using enrichment methods.

Steroid and other transformations using spores or vegetative cells

Biochemical activity on the part of fungal spores was first observed in 1958 when Gehring and Knight noticed that a suspension of spores of *Penicillium roqueforti* could convert caprylic acid (octanoic acid) to 2-heptanone by the process illustrated in Figure 3.12.

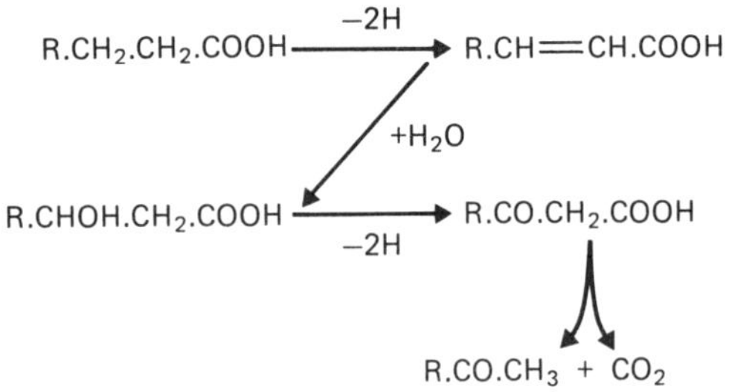

$$R.CH_2.CH_2.COOH \xrightarrow{-2H} R.CH{=}CH.COOH$$

$$+H_2O$$

$$R.CHOH.CH_2.COOH \xrightarrow{-2H} R.CO.CH_2.COOH$$

$$R.CO.CH_3 + CO_2$$

Figure 3.12   Process of ketone formation from fatty acids by spores of *Penicillium*

Definition and role of steroids

In 1962 Knight obtained the 11 α-hydroxyl derivative of progesterone using conidia of *Aspergillus ochraceus* (Figure 3.13). The possibility thus arose that spores might be used for useful bioconversions of steroids. Steroids are compounds having a common skeleton in the form of perhydro-1,2-cyclo-pentano-phenanthrene (Figure 3.14).

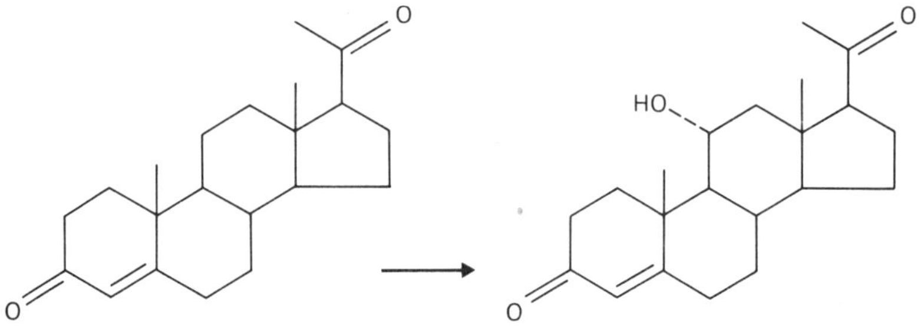

Figure 3.13   11α-hydroxylation of progesterone by *Aspergillus ochraceus*

There are microbial steroids (ergosterol), plant (diosgenin) and animal steroids (cholesterol, bile acids, corticosteroids and sex hormones). Steroids are involved in the regulation of metabolism and certain of them have important uses as therapeutic agents.

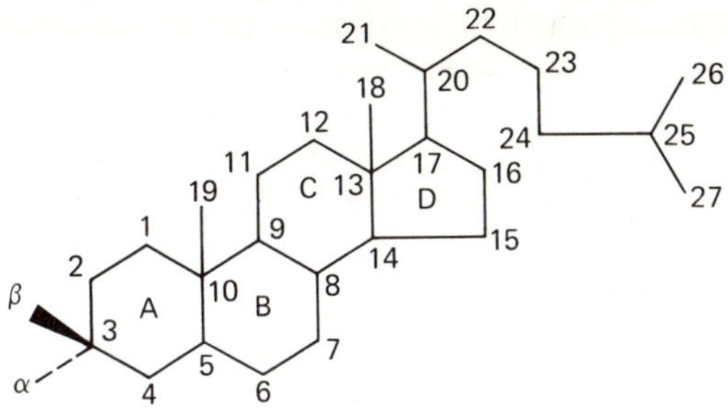

Figure 3.14   Skeleton of the steroids

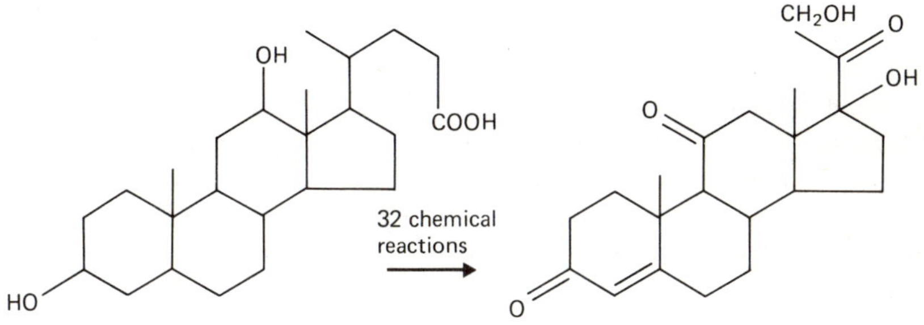

desoxycholic acid

cortisone

Figure 3.15   Synthesis of cortisone from desoxycholic acid

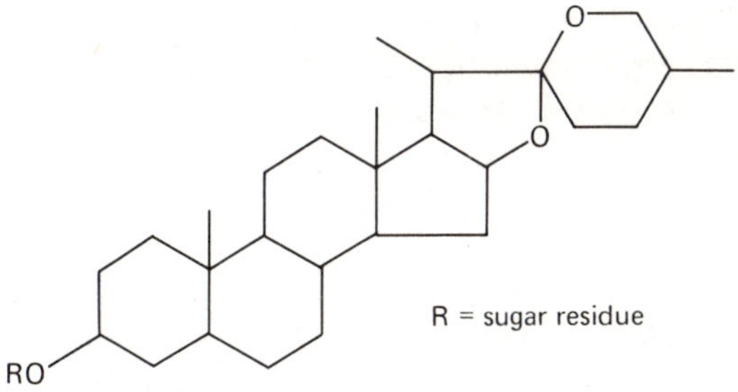

Figure 3.16   Structure of diosgenin

The beneficial anti-inflammatory effect of cortisone in, for example, the treatment of rheumatoid arthritis and skin diseases, is associated with the presence of an oxygen function at carbon atom 11. To obtain cortisone by chemical synthesis from desoxycholic acid requires 32 different chemical steps, of which 12 are involved in moving an oxygen function from position 12 to 11 with a low overall yield (1 kg cortisone was obtained from 500 kg of desoxycholic acid. Figure 3.15).

In 1952 Peterson, a research worker at the pharmaceutical company Upjohn, discovered a strain of *Rhizopus nigricans* which gave 11α-hydroxy-progesterone from progesterone with an 80% yield. Progesterone itself is readily obtained economically by chemical synthesis from diosgenin which is an abundant plant saponin (Figure 3.16).

Industrial bioconversions

This discovery opened the way to the use of steroid bioconversions by the pharmaceutical industry. Among the bioconversions used on an industrial scale, two groups of reactions may be distinguished:

(1) Hydroxylations These may take place at various positions:

11α-hydroxylation. 11α-hydroxyprogesterone can be obtained from progesterone which itself may be produced by chemical synthesis or from diosgenin. 11α-hydroxyprogesterone was formerly a principal intermediate in the preparation of cortisone and numerous other steroids. An 11-hydroxylation of the synthetic derivative known as Reichstein's Compound S yields epicortisol, which can be converted chemically into the very active 9α-fluoro derivative.

11β-hydroxylation. The natural steroid hydrocortisone (cortisol) can also be obtained from Reichstein's Compound S, either by direct hydroxylation using *Curvularia lunata* or a mammalian enzyme preparation, or indirectly by catalytic hydrogenation of cortisone and its derivatives. Because of the stereochemical properties of the family of steroids containing cortisol and cortisone, direct introduction of a hydroxyl group into the 11β position is difficult and yields are low. However, for the same stereochemical reasons, catalytic hydrogeneration of cortisone derivatives is stereospecific giving a high yield of the 11β hydroxy compounds. The usual route then to the 11β hydroxy compounds is microbiological introduction of an 11α hydroxyl group, chemical oxidation to the 11-keto compound and catalytic hydrogenation to the 11β hydroxyl derivative. Cortisol is an intermediate in the production of the 9α fluoro derivatives which are highly effective anti-inflammatory agents.

16α-hydroxylation. Direct hydroxylation in this position is carried out readily by a number of species of *Streptomyces*. These reactions became important when it was appreciated that the 16-hydroxy derivative of 9α-fluoroprednisone is a powerful anti-inflammatory agent which is free of certain side effects of 9α-fluoroprednisone itself.

21-hydroxylation. Although hydroxylation in the 21 position is relatively easy by chemical processes, and microorganisms are not widely used for this reaction in industry, the ability of some fungi to introduce a 21-hydroxy constituent may be useful in a research laboratory. *Aspergillus niger* and *Ophiobolus herpotrichus* are two species of fungi able to carry out this transformation converting, for example, progesterone into desoxycortisone.

(2) Dehydrogenation
Arthrobacter simplex is one of several microorganisms which can bring about 1:2 dehydrogenation of several steroids. Prednisolone is obtained thus from cortisone and is then used for the manufacture of further useful derivatives. It is also possible to effect the simultaneous dehydrogenation and removal of the side chain of progesterone to give androstadienedione using *Streptomyces lavendulae.* Pyrolysis of this compound leads to the sex hormone oestrone.

Bioconversion methods

In the direct method, the chosen microorganism is cultured in a medium containing only the minimum nutrients required for growth (which facilitates subsequent purification of the steroid). At the end of the growth period the steroid to be converted is added dissolved in a small quantity of a water miscible solvent such as ethanol, acetone or propylene glycol. Most steroids have a low solubility in water and thus precipitate out as a fine colloidal suspension in the medium. Because of the relatively high toxicity to micro-organisms of the solvents used it is usually only possible to add 0.5–2 g/l of the steroid to be altered. The transformation takes some 20–40 hours, and is followed by extraction of the new steroid with water-immiscible solvents such as chloroform, and chromatographic analysis of the product.

Another type of process also involves preliminary growth of the chosen microorganism, this time on a rich medium to allow the production of a high yield of biomass. This is then filtered, washed, resuspended in distilled water and the steroid precursor added in an appropriate solvent. In an aerated vessel, with efficient stirring and maintained at the optimum temperature, good yields of transformation products can be obtained in a few hours; the high yield is an advantage in extraction.

A further possibility is to use polyacrylamide gel columns bearing an entrapped fungal mycelium (e.g. *Curvularia lunata*) or a previously purified enzyme. A solution of the steroid to be converted is then passed through the column: transformation is thus continuous and the prepared column may be regarded as a stationary enzyme system.

Preparation of spores

When using fungal or actinomycete spores for bioconversions, it is essential to obtain them in large quantity. The optimum conditions for sporulation have to be determined for each strain used. Frequently, spores are best produced by surface culture on agar media or on layers of bran or cereals soaked with water, distributed in Roux bottles or Fernbach flasks.

The water activity of the medium and relative humidity of the gas phase play an important role in determining the yield of spores (Table 3.10).

Table 3.10 The influence of water activity on spore production by several fungi (results expressed as 10^{11} conidia per flask)

ml *water* *per flask*	*Aspergillus* *ochraceus* NRRL 405	*Aspergillus* *niger* ATCC 9142	*Mucor* *griseocyanus* ATCC 1207A	*Penicillium* *chrysogenum* WIS 53–414
40	< 1	+	—	+
60	3·2	2·1	—	1·8
80	3·8	2·0	–	2·0
100	3·0	+	–	1·0
120	+	—	6·3	+
140	–	–	4·5	–
160	–	–	–	–

(Fernbach flasks of 2·8 l capacity containing 200 g barley were sterilized for 1 h at 121°C. Incubation was at 28°C for 7 days in the case of *A. ochraceus* and at 25°C for 14 days for other fungi. The relative humidity was 50–60%)

The spores are harvested by adding water to the flasks, shaking vigorously and filtering the resulting suspension through glass wool to remove mycelium and medium fragments; the filtrate contains the spores, which are centrifuged and washed.

High yields of spores are also obtainable in submerged culture but the composition of the medium is very important and varies from strain to strain (Table 3.11).

Streptomyces spore preparations can likewise be obtained in submerged culture.

The spore crop can either be used within a short period after harvesting or preserved by refrigeration at about −20°C. Spores of some species (e.g. *Aspergillus ochraceus*) remain viable under refrigeration for long periods: with others (e.g. *Septomyxa affinis*) viability is lost after about eight weeks.

Bioconversion

The spores are resuspended in a buffer solution (phosphate, acetate or citrate) in which they do not germinate. The steroid is added, either dissolved in a water-miscible solvent or as a microcrystalline suspension in 0.01% Tween 80

Table 3.11 Sporulation of filamentous fungi in submerged culture
(number of spores expressed per ml after
5 days incubation)

Species	Medium	Number of spores
Mucor griseocyanus	A	small
Aspergillus ochraceus	B	2×10^8
Curvularia lunata	B	$<1 \times 10^7$
Fusarium equiseti	D	$6{\cdot}4 \times 10^7$
Fusarium oxysporum	D	$2{\cdot}5 \times 10^8$
Trichoderma viride	C	$5{\cdot}0 \times 10^7$

(In the above table Medium A contains malt extract, B contains
corn steep liquor and molasses, C glucose and mineral salts and D
sucrose and mineral salts. If the level of zinc in Medium D is reduced
to $2000\,\mu g/l$ spore production is low.)

solution. The final concentration of the steroid which can be added depends on
the activity of the spores: for example, in the $1:2$ dehydrogenation of
Reichstein's Compound S by spores of *Septomyxa affinis*, $5\,g/l$ of the
compound is appropriate. Sometimes the transformation occurs only in the
presence of sugars: hydroxylation of steroids by spores of *Aspergillus
ochraceus* or *Mucor griseocyanus* requires a concentration of between
$0{\cdot}2$–$0{\cdot}4\%$ glucose.

Steroid transformations may be inexacting as regards temperature and pH
conditions; thus, hydroxylation of progesterone occurs within the pH range
4–$8{\cdot}5$ and temperature range $23°C$ to $34°C$. After reaction, the spores are
separated by filtration. The transformation product may either be present in
the filtrate, or bound to the spores. In some instances the spores can be reused.

The use of spores for bioconversion has a number of advantages. Sterilization of the suspending medium is not necessary, small concentrations of
tetracycline, neomycin or other antibiotics being sufficient to inhibit the
growth of bacterial contaminants. Neither is it necessary to sterilize the air
needed for oxygen requiring processes, e.g. dehydrogenations. Spores are
frequently 3–10 times more active, and may be stored under refrigeration for
much longer periods than mycelium. Active suspensions can be prepared in
simple buffers, obviating foam problems and simplifying solvent extraction.
Using a 1000 litre fermenter it is possible to transform 1 kg of substrate per
batch. Table 3.12 indicates some of the bioconversions which are possible.

Transformations of compounds other than steroids can also be carried out
with fungal spores. Thus starch can be saccharified using spores of *Aspergillus
wentii* and glucose converted to mannitol (75% yield) by conidia of *Aspergillus
candidus*. The spores of a strain of *Fusarium* can be used to convert penicillin V
to 6-aminopenicillanic acid, an important intermediate in the production of

synthetic penicillins. These few examples illustrate the wide range of possibilities using the metabolic activities of fungal spores.

Table 3.12 Bioconversions of steroids using spore suspensions

Substrate	Reaction	Conidia per ml	Organism	Incubation period (h)
Progesterone	11α-hydroxylation	2×10^8	*Aspergillus ochraceus*	16 (100% conversion)
Reichstein's Compound S	11α-hydroxylation	2×10^8	*Aspergillus ochraceus*	48 (100% conversion)
Progesterone	1:2 dehydrogenation with side-chain degradation	—	*Septomyxa affinis*	—
19-nortestosterone	1:2 dehydrogenation without side-chain elimination	—	*Septomyxa affinis*	—
Oestrone	15α-hydroxylation	$1{\cdot}6 \times 10^6$	*Fusarium moniliforme*	—
Oestradiol	17-oxidation	2×10^8	*Streptomyces diastaticus*	—

3.3   Extraction of the desired product

Most often, the desired fermentation product is released in low concentration into the aqueous culture medium: extraction is then a matter of separating it from the cells and dissolved and suspended media constituents. Occasionally the product is contained within the microbial cells which must therefore be first concentrated and then disrupted.

The initial culture medium contains dissolved nutrients of low molecular weight (sugars and mineral salts) and/or higher molecular weight (peptides and vitamins), sometimes together with suspended particulate materials such as soya meal. Ideally, at the end of the fermentation all media nutrients should have been completely used and only water, microbial biomass and released metabolites (including the desired product) should remain: needless to say, this ideal is rarely if ever attained.

The molecular weight of the products may be low (lactic and glutamic acids) medium (antibiotics) or high (enzymes); their concentration may be as low as a few mg per litre (as in the case of vitamin B_{12}) or as high as 130 g/l (in the case of lactic acid) or even 300 g/l (formation of L-sorbose by *Gluconobacter oxydans* subsp. *suboxydans*). Some indication of the range of concentrations of nutrients and products is given in Table 3.13.

Table 3.13 Some nutrients and products in
industrial-scale fermentations

Substance	Molecular weight	Specimen values for concentration (g/l)
Nutrients		
Glucose	180	50
Urea	60	5
$(NH_4)_2SO_4$	132	2
KH_2PO_4	136	1
Dried yeast	—	5
Soya meal	—	10
Products		
α-amylase	48,900	20
Vitamin B_{12}	1375	0.023
Streptomycin sulphate	582	4
Sodium glutamate	169	35
Lactic acid	90	130

The various stages in the extraction and recovery of a soluble extracellular
product are shown in generalized form in Figure 3.17.

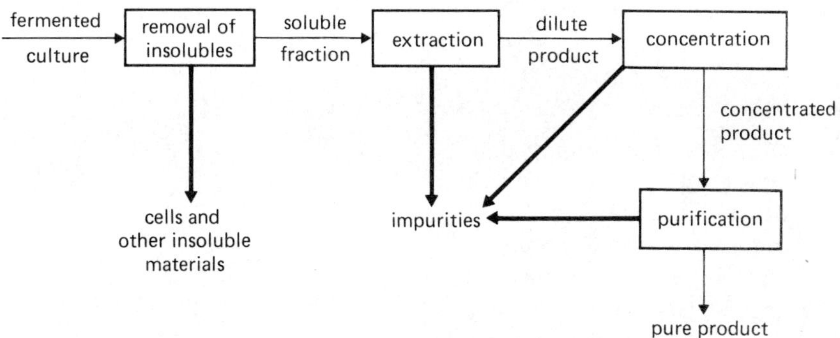

Figure 3.17   Generalized scheme for the recovery of a soluble, extracellular
product

Removal of insolubles

Insoluble material, comprising mainly microbial cells, may represent some
10%, or more, of the culture and is separated first. The harvested biomass may
itself be of commercial value, for example as an animal feed additive; it may

contain some or even all of the desired product in an intracellular form. Figure 3.18 indicates, as a classification, some of the methods usually employed for removing insolubles.

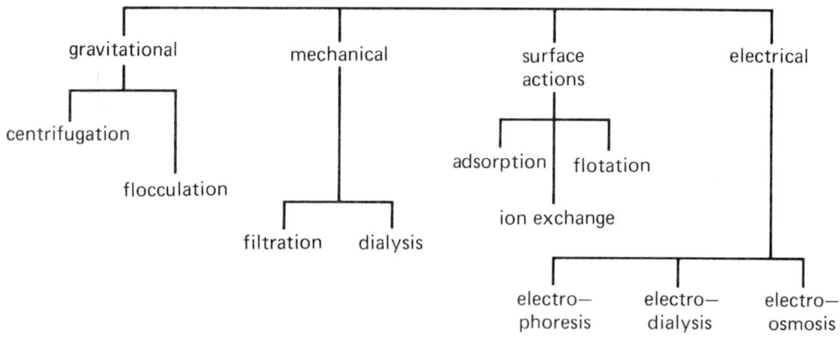

Figure 3.18   Principal methods for the removal of insolubles

Centrifugation

Recovery or removal of unicellular microorganisms such as yeasts and bacteria can be effected by centrifugation. According to Stokes's law, the rate of sedimentation of spherical particles in a Newtonian fluid is proportional to the square of the diameter of the particles (in this case, microbial cells).

$$V = \frac{(\rho_p - \rho_L)d^2\omega^2 r}{18\mu}$$

V = rate of sedimentation (cm/s)
ρ_p = density of particles (g/cm^3)
ρ_L = density of liquid (g/cm^3)
d = diameter of particles (cm)
μ = viscosity of liquid (g/cm/s)
ω = angular velocity (rad/s)
r = distance of particles from the axis of rotation.

When continuous centrifugation is employed the flow Q through the apparatus is described by the following equation:

$$Q = \frac{(\rho_p - \rho_L)d^2\omega^2 r V}{18\mu S_e}$$

Q = flow rate through the apparatus (cm^3/s)
V = volume of liquid within the apparatus (cm^3)
S_e = effective distance travelled by the sedimenting particle (cm).

It can thus be seen that the factors influencing rate of deposition during centrifugation are the difference in density between the cells and liquid, the diameter of the cells, and viscosity of the liquid. Table 3.14 gives the size range of certain groups of microorganisms and the appropriate centrifugation method.

Table 3.14 Dimensions of microorganisms and centrifugation methods

Microorganism	Diameter (μ)	Centrifugation method
Viruses, phages	0·01–0·1	ultracentrifuge
Bacteria	0·3–3·0	normal
Yeasts	4·0–7·0	normal
Filamentous fungi	10·0–150	normal

Since the mean dimension of most bacteria is approximately one-fifth that of yeasts, a centrifugal force some twenty-five times greater is needed to separate them. This is an important consideration in choosing strains for large scale cultivation as food protein supplements and for industrial fermentations generally.

The centrifuge has a further application in the separation of partly emulsified liquids of different density. For example, when organic solvents are used to extract and purify compounds found in the culture liquor, rapid separation of the solvent and aqueous phases is usually effected by passing the mixture through a continuous centrifuge.

Flocculation

Flocculation depends on the fact that clumps of microbial cells, though having the same density as individual cells, sediment more rapidly because of their increased dimension (see preceding section). Production of floccules is usually brought about by neutralization of electrical charges on the cell surface. The cells of bacteria and yeasts, when suspended in a medium of neutral pH, carry on overall negative charge because of the presence of carboxyl and phosphate groups on their walls. Flocculation is influenced by the characteristics of the cell wall, the ionic environment, the pH of the medium and the flocculating agent used. Properties desirable in a flocculating agent include:

(1) rapid reaction with the cells to be flocculated
(2) absence of toxicity
(3) minimal alteration of cell constituents or products

(4) absence of marked acidity or alkalinity, and ability
   to act at near-neutral pH
(5) low cost and/or high activity at low concentration
(6) ease of separation from the final product.

Flocculation is an essential stage in the brewing of beer. Ale yeasts have the property, apparently controlled by a dominant gene, of flocculating spontaneously towards the end of the fermentation, provided calcium ions are present. (Urea disperses the floccules indicating that hydrogen bonding plays a part in their formation.) It is thought that flocculation is brought about by the formation of salts between calcium ions and the carboxyl groups on the surface of adjoining cells; the floccules thus formed are then further stabilized by hydrogen bonding.

Flocculating agents which act by neutralizing the negative charge present on the cell surface include aluminium sulphate ($0\cdot1-0\cdot5\%$ final concentration), calcium chloride ($0\cdot1-0\cdot5\%$) and titanium tetrachloride ($0\cdot01-0\cdot02\%$). Cationic detergents, such as quaternary ammonium compounds, alkylamines and alkylpyridinium salts, are also used at $0\cdot01-1\cdot0\%$ concentration.

Bacteria may be flocculated without charge neutralization by certain synthetic anionic or non-ionic polyelectrolytes (polyacrylamide, polystyrene sulphate, polyglutamic acid). This effect is due to the interaction of the polymer molecule with the surfaces of several neighbouring cells. It is likely that synthetic polyelectrolytes will, in the future, find an increasing use in the cheap recovery of microbial cells. Hydrophilic polymers such as gelatin, carboxymethyl cellulose, alginates and dextrans can also induce flocculation.

Filtration

This process is mostly used for the separation of mycelial fungi or actinomycetes from the culture fluid. Filter presses may be used, but the common method is by rotary vacuum drum filtration. The slenderness of actinomycete filaments may cause filter blockage and necessitates the addition of a filter aid, such as diatomaceous earth, perlite, asbestos or cellulose pulp. The filter aid becomes deposited over the surface of the filter forming, together with the retained filaments, a 'filter cake'. In rotary vacuum filtration, the culture is drawn into a horizontal, fabric-covered, perforated rotating drum, the interior of which is under a slight vacuum. The filter cake collecting on the outside of the drum is continuously scraped off by a fixed blade called a 'doctor'. Different types of filtration are summarized in Table 3.15.

Filtration is cheaper than centrifugation. However, in the case of bacterial processes, it can usually replace centrifugation only if the culture is given some preliminary treatment to increase filtrability. This may involve, for example, acidifying followed by moderate heat treatment to coagulate proteins. Table 3.16 gives the results of a laboratory scale experiment illustrating this point.

Table 3.15 Filtration data for selected fermentations

Filtration data	Fermentation product				
	cortisone	kanamycin	neomycin	penicillin	streptomycin
Type of filtration	pressure	vacuum + filter aid	vacuum + filter aid	vacuum + filter aid	vacuum + filter aid
Filtration rate (gal/h/ft^2)	40·8	2·1	3·2	35–45	3–20
(l/h/m^2)	1998	103	155	1715–2200	145–970
% dry weight in fermented culture	20	7	2–8	2–8	2–6
vacuum (in Hg)	10–20	20	18–20	20	20
% Water in deposit	38–45	—	—	60–70	—

Table 3.16 Effect of various pre-treatments on the filtration rate of a culture of *Acinetobacter calcoaceticus*

Acidification	Heat	pH at time of filtration	Filtration temperature (°C)	Filtration rate (ml/min)
—	—	7·0	25	0·53
+	—	3·5	25	2·22
+	+	3·5	85	10·00
—	+	7·0	85	0·16

Filtration can also be used to sterilize certain soluble products and, occasionally, culture media. Press or membrane filters may be used, the latter (membranes of cellulose acetate or other cellulose esters) being applicable only where the liquid is non-viscous and contains little particulate matter.

Flotation

This process depends on the tendency of particles (e.g. microbial cells) in a liquid suspension to attach to gas bubbles. A gas, normally air, is injected into the bottom of the fermenter; the bubbles rise, with their attached particles, to form a surface foam which can be removed by skimming (Figure 3.19). This process has been called micro-flotation. It may be necessary to add foam inducing compounds, particularly if an antifoam agent was added to the culture during fermentation.

The foam produced should have a high gas:liquid ratio so that, when broken, it will yield a suspension with a high concentration of microbial cells. The ratio of the concentration of cells in the broken foam to that in the original

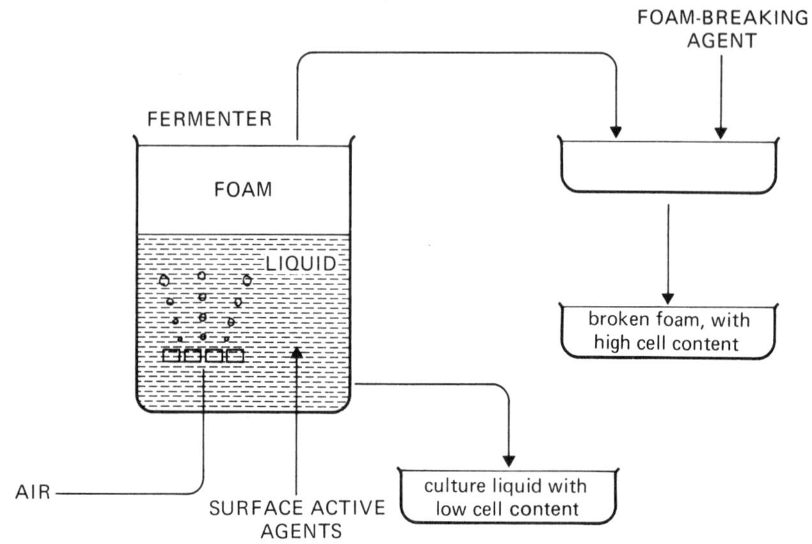

Figure 3.19   Generalized scheme for separation of cells by flotation

suspension is referred to as the concentration factor. A similar expression, sometimes used, is the enrichment ratio, the ratio of the number of cells per unit volume of the fermented liquor before and after foam separation. Both ratios are improved by the addition of substances favouring fixation of particles to the bubbles (ionic amides, cetylpyridinium chloride, mineral salts or cationic detergents).

In another process an immiscible phase, usually consisting of long chain fatty acids or amines, is added. Less gas is required than in micro-flotation and the cells fixed to the bubbles form a stiffer foam at the surface. This process may be called sublation (= act of lifting up, taking or carrying away).

Flotation methods can be highly efficient: for example, with cultures of *Escherichia coli* plus a cationic detergent (ethyl hexadecyl methyl ammonium bromide, 0·04 mg/ml) the enrichment ratio can be of the order of 10^6. In the case of some *Bacillus* species, the spores can be concentrated by flotation with secondary amines and the vegetative cells recovered by flotation with fatty acids. Filamentous algae can often be harvested by flotation without recourse to surface-active agents, because entrapped O_2 bubbles produced during photosynthesis cause the whole growth mass to rise. Laboratory studies suggest that flotation methods could be applied to a wide range of micro-organisms.

Other methods

Small unicellular microorganisms and viruses can be separated from culture fluids by adsorption onto ion-exchange resins, from which they can subsequently be detached by appropriate reagents. Disadvantages of this method include the large quantities of resin required and the susceptibility of the resins to biodeterioration caused by growth of unwanted microorganisms (e.g. proteolytic organisms) which might attack or contaminate fermentation products subsequently passed through the column.

Two phase liquid separation of cells has been investigated experimentally. The distribution of microbial cells between two phases is as follows:

$$\frac{C_1}{C_2} = \exp\left[\frac{M\lambda}{RT}\right]$$

C_1 = concentration of cells in phase 1
C_2 = concentration of cells in phase 2
M = surface area of cells
λ = constant for each two phase system and each type of cell
R = gas constant
T = temperature (°K).

Using such a system, it has proved possible to separate the spores of *Bacillus thuringiensis* from the cytoplasmic protein crystals of the organism (this crystalline protein is toxic to certain moths which attack tobacco plants and thus has potentiality as a specific insecticide). By two phase liquid/liquid separation, using water and carbon tetrachloride, 99% of the crystals with only 1% of the spores can be concentrated in the aqueous phase; the remaining spores separate into the CCl_4 phase.

Two phase liquid/liquid separation is also extensively employed in the concentration and purification of viruses for use as vaccines. Fluorocarbon compounds (Freons) often form the organic phase.

Microorganisms, including viruses, carry an overall electric charge, and thus migrate in an electric field (Table 3.17); electrophoresis is therefore a

Table 3.17 Electrophoretic mobility of selected microorganisms

Organism	Suspending fluid	pH	mobility (μ/s/V/cm)
Streptococcus pneumoniae	0·04 M phosphate	7·3	4·2
Staphylococcus aureus	0·013 M phosphate	7·4	1·8–2·0
Pseudomonas sp.	0·01 M phosphate	6·9	2·0–2·1
Brucella abortus	0·01 M phosphate	7·5	3·72
Escherichia coli phage	0·08 M NaCl	5·78	0·43

potential means of separating organisms from culture fluid. It is likewise theoretically possible to separate, by electrophoresis, different kinds of cells; for example, ascospores of *Saccharomyces cerevisiae* have been separated from diploid cells of the organism using electrophoresis.

Extraction, concentration and purification of soluble products present in the culture liquid

When, as is frequently the case, the desired product is dissolved in the liquid phase of the culture, the problem (having removed cells and other solids) is to find a convenient means of obtaining this product in a concentrated form, freed of media constituents and unimportant products of the fermentation. Several general methods will be described.

Solvent extraction

Choice of solvent depends on the product to be extracted. A substance dissolved in the culture fluid will, on addition of an organic solvent, become distributed between the solvent and aqueous phases according to a partition coefficient k:

$$k = \frac{C_s}{C_e}$$

C_s = concentration in the solvent phase
C_e = concentration in the aqueous phase.

The value of this partition coefficient varies with pH which must be adjusted to ensure the most efficient extraction. To give an examples of solvent extraction, penicillin is an organic acid with the formula shown in Figure 6.1.

The sodium salt is highly soluble in water whereas the free acid is much more soluble in organic solvents e.g. amyl acetate. Thus, during solvent extraction at a low pH penicillin is almost entirely present as the undissociated acid, and passes into the solvent phase. To carry out the extraction, amyl acetate is intimately mixed with the filtered culture fluid, together with sufficient sulphuric acid to reduce the pH to 2·0; penicillin, as the free acid, passes into the solvent phase. The resulting emulsion is then separated by centrifugation. The solvent phase is mixed with 1/20 volume of sodium bicarbonate buffer at pH 8·0. Penicillin then passes into the aqueous phase as the sodium salt at a concentration 50–100 times greater than that present in the original culture fluid, and considerably purified.

The organic solvents employed (alcohols, esters, ketones, ethers, hydrocarbons) are usually both costly and inflammable. Their use on an industrial scale always therefore involves both solvent re-distillation and stringent safety measures to minimize risk of fire and explosion.

Adsorption and elution

The compound to be separated, usually present in low concentrations in aqueous solution, is adsorbed onto a suitable solid, usually in column form, and then eluted with a much smaller volume of buffer at an appropriate pH. Columns of aluminium hydroxide, cellulose powder, activated charcoal, silica or ion exchange resins may be used.

As an example, in purification of the basic antibiotic, streptomycin, filtered culture medium is passed through a column of an anionic resin in the form of the sodium salt; streptomycin displaces sodium ions from the resin:

$$\text{R-COO}^-\text{Na}^+ + \text{streptomycin} \rightarrow \text{R-COO}^-\ \text{streptomycin}^+ + \text{NaOH}$$

The bound streptomycin is then eluted from the charged resin by passing a mineral acid, such as $0{\cdot}2\,\text{M}\,\text{HCl}$, through the column:

$$\text{R-COO}^-\ \text{streptomycin}^+ + \text{HCl} \rightarrow \text{R-COOH} + \text{streptomycin}^+\ \text{Cl}^-$$

The column is finally regenerated by passing an NaOH solution through it:

$$\text{R-COOH} + \text{NaOH} \rightarrow \text{R-COO}^-\ \text{Na}^+ + \text{H}_2\text{O}$$

The adsorption capacity of an appropriate resin is of the order of 200–300 g of the antibiotic per litre of resin. In practice, a series of resin columns is used, the culture filtrate being passed through each column in succession. When the first column is saturated it is removed and eluted, a fresh column is added to the end of the series, and so on. The used columns are regenerated by NaOH treatment and are then ready for reuse.

The antifungal antibiotic, actidione, is produced, together with streptomycin, by certain strains of *Streptomyces griseus*. To separate the two antibiotics, the culture filtrate is passed first through columns of activated charcoal. Actidione is adsorbed and is subsequently eluted with acetone; streptomycin passes through and is then concentrated by adsorption onto an ion exchange resin.

Precipitation

Some products can be precipitated directly from the culture filtrate, either by adding a water-miscible organic solvent, or by adding an agent which causes formation of insoluble complexes or salts. Commercial production of dextran (6-glucopyranose polysaccharides) affords an example of the first method. Extracellular dextrans are formed by *Leuconostoc mesenteroides*, using sucrose as the substrate; at the end of the fermentation the culture is mixed with an equal volume of methanol to precipitate the polysaccharide. This precipitate is allowed to sediment, leaving medium constituents and cells in the supernatant, which is removed. The precipitated dextrans are purified by dissolving them in a small volume of water and reprecipitating with methanol; the process is repeated to yield an increasingly pure product.

Examples of precipitation as insoluble complexes are found in the isolation of certain antibiotics. Thus, in the isolation of tetracycline use may be made of the fact that it forms an insoluble complex with certain quaternary ammonium salts. In the production of penicillin a precipitation stage may be used as one of the final purification processes. The organic base procaine is added to the concentrated aqueous solution resulting from solvent extraction and the crystalline precipitate of the procaine salt of penicillin is filtered and washed: it may then be transformed into the sodium or potassium salt, but the procaine salt also has a medical use since, after intramuscular injection it slowly releases penicillin and thus acts as a medium long acting antibiotic.

A slightly different example of precipitation is found in the isolation of riboflavin. This B-group vitamin is released into the culture medium by various organisms including *Ashbya gossypii*. It can be reversibly reduced to an insoluble form by adding sulphites or stannous salts to the culture or by inoculating it with *Streptococcus faecalis*.

Liquid ion exchange

Solvent extraction has the advantage that it can be operated as a continuous, countercurrent, process; it is also fast, an important consideration when the product is unstable. However, on a large scale, ion exchange resins are more economical because the cost of regenerating them is less than the cost of recovering solvents. Ion exchange columns cannot be employed in a strictly continuous manner and they become blocked if insoluble materials are present in the culture fluids.

Liquid ion exchange techniques combine some of the advantages of both methods. The principle is shown in Figure 3.20.

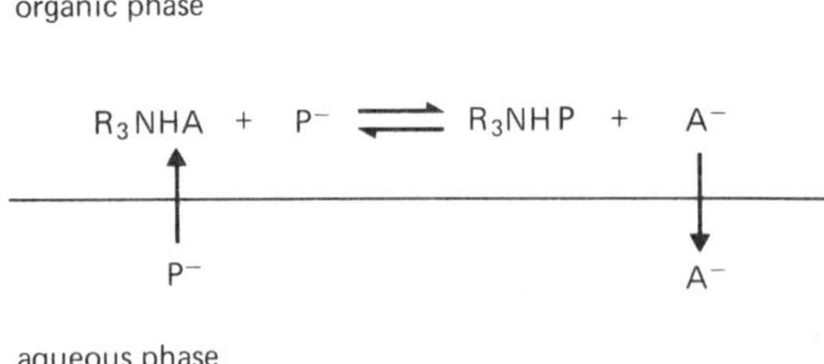

Figure 3.20   Use of liquid ion exchange to extract an anion P from a culture fluid (aqueous phase)

An ion exchange liquid consists of a solution of the salt of an organic amine R_3N and an acid HA, in an organic solvent. When this liquid is intimately mixed with an aqueous solution of an anion P^-, ion exchange occurs, the product displacing the anion of the acid, which passes into the aqueous phase. The two phases are separated by centrifugation and the product recovered from the organic phase, either by precipitation, or by a further ion exchange which transfers it back to an aqueous solution.

This technique, successfully used in the extraction of uranium from its ores, has been tested on a pilot plant scale for the extraction and purification of terramycin: at pH9·5–10·2, iso octyl alcohol containing 5% cetylpyridinium bromide gave 95–97% recovery. The most widely used liquid anion exchange compounds are organic amines with molecular weights in the range 250–600, and a variety of liquid cation exchange compounds (Table 3.18). The organic solvents used include benzene, hexane, chloroform and various aliphatic alcohols.

Gel filtration

This process affords a means of separating molecules of different sizes. The gels (dextrans, polyacrylamides or polystyrene derivatives) are dispersed in an aqueous phase, forming a three-dimensional mesh. Larger molecules diffuse into the gel mesh more slowly than do small ones. Thus, when a charged column of gel is eluted by passing fresh aqueous phase through it, larger molecules, lying in the interstitial spaces, are more readily washed out and appear in the first effluent samples. This technique has been used for the removal of traces of protein from crude benzyl penicillin. Such proteinaceous impurities have been considered to be, at least partly, responsible for the allergic response which some individuals suffer following repeated treatment with penicillin. Some data are presented in Table 3.19.

Table 3.18 Liquid ion exchange compounds

Compound	Manufacturer's name	Manufacturer
Anion exchangers		
Triisooctylamine	TIOA	Carbide, Gulf
Trilaurylamine	TLA	ADM Armour
Trialkylmethylamine	Primene JM-T	Rohm and Hass
1(3-ethylpentyl)-4-ethyloctylamine	Amine 21 F 81	Union Carbide
Cation exchangers		
Di(2-ethylhexyl) phosphoric acid	D 2 EHPA	Union Carbide
Dodecyl phosphoric acid	DDPA	Dow Chemicals
Tri-*n*-octylphosphine oxide	TOPO	Eastman Kodak

Table 3.19   Purification of benzyl penicillin by gel filtration; process details

Filter	Sephamatic GF 08-10 (volume 500 l)
Gel	Sephadex G-25
Charge	50 l of 40% solution per cycle
Eluant	500 l per cycle
Product obtained	375 l purified benzyl penicillin 98% yield
Duration of cycle	110 minutes
Capacity	27 l per hour of impure solution

Gel filtration is also used in the purification of tetanus and diphtheria toxins. Columns of 13 litre capacity are used in series: a Sephadex G-100 column first removes low molecular weight substances: a second column, of Sephadex G-200, serves to fractionate high molecular weight compounds, allowing a high degree of purification of the toxin.

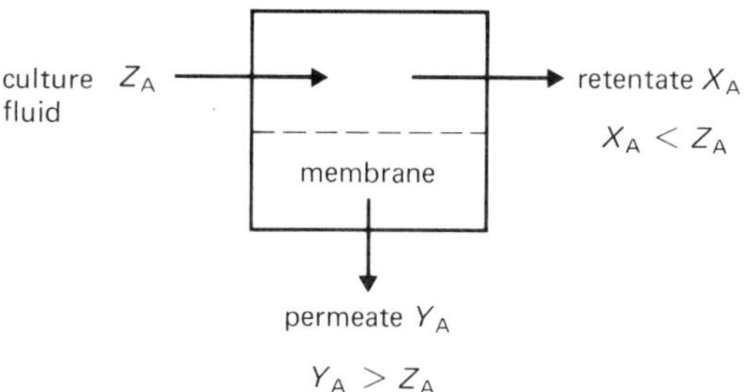

X_A, Y_A and Z_A are the concentrations of A in retentate, permeate and culture fluids respectively

Figure 3.21   Simplified scheme for the separation of a substance A using membranes

Membrane methods

Figure 3.21 represents a general scheme for membrane separation. Compound A, present in the culture liquid, preferentially passes across the membrane; after passage through the system, the retentate contains a lower concentration and the permeate a higher concentration of compound A.

Membranes may be used in a number of different ways for the isolation and purification of fermentation products (Figure 3.22).

In *dialysis*, solute molecules pass across a separating membrane from a concentrated solution to a more dilute one, this process being particularly useful for the removal of salts from solutions of proteins, which do not pass dialyzing membranes.

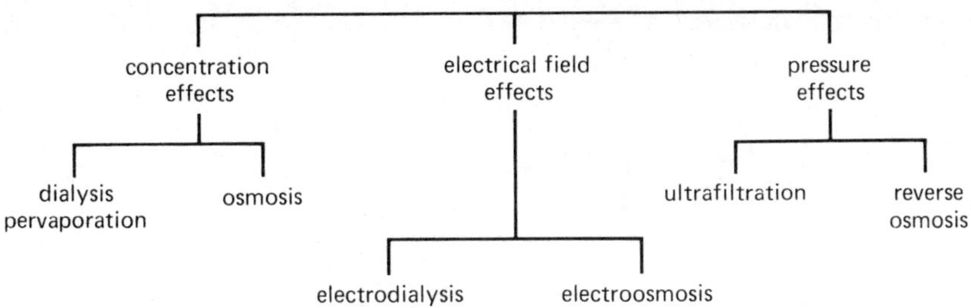

Figure 3.22   Classification of membrane separation methods

In *electrodialysis*, the passage of electrically charged solute molecules across the membrane is facilitated by applying an electrical field.

In *osmosis* it is the solvent which passes across the membrane, from a weak solution into a more concentrated one, the membrane being impermeable to the solute. It is again possible to accelerate transfer across the membrane by applying an electrical field, the process than being called *electroosmosis*.

In the process termed pervaporation, a liquid phase is present on one side of the membrane and a gaseous phase on the other. The volatile constituents of the liquid phase pass across the membrane, and evaporate into the gas phase, the latter may be held at a reduced pressure to accelerate the transfer.

When high pressure is applied to a solution in order to force solvent across the membrane from a strong solution to a weaker one (i.e. against an osmotic gradient) the process is referred to as *reverse osmosis*. If the application of pressure results in the passage across the membrane of low molecular weight solutes as well as solvent, the process is called *ultrafiltration*. Ultrafiltration is essentially a method of concentrating high molecular weight solutes; because of the high pressures which have to be used (100–1000 pounds/in^2 [70,000–700,000 kg/m^2]), the membranes have only a limited life.

Endocellular products

Consideration has so far been given to the excretion of soluble substances occurring in the culture filtrate. However, the desired product may well be partly or wholly intracellular (antibiotics, for example, may be partly intracellular). Sometimes it is possible to extract such products from the whole culture by modifying the pH or by using solvents such as chloroform.

When the intracellular product is less stable (e.g. enzymes) other methods of liberation have to be used. These include the use of enzymes to destroy the cell wall, thermal shock (vitamin B_{12}), osmotic shock, alternate freezing and thawing, grinding of frozen mycelium at a temperature of $-30°C$, bursting of

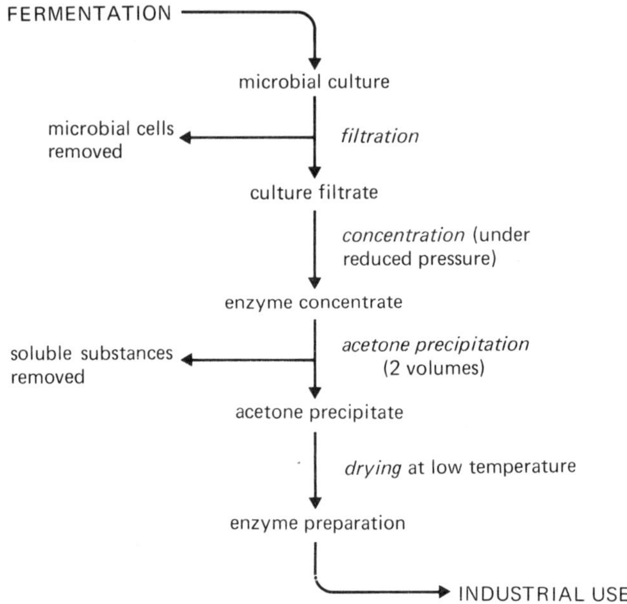

Figure 3.23   Simplified scheme for preparation of a crude enzyme (extracellular)

cells by shearing in high pressure extrusion dies, ultrasonic disintegration and high-speed milling with glass beads.

Subsequent extraction procedures vary widely according to product, and the degree of purity required. Figures 3.23 and 3.24 compare the comparatively simple procedures for isolation of an extracellular enzyme with the more lengthy one which would be needed for extracting an intracellular enzyme. When the product is for medical use, particularly if it is to be used for injection, the extraction and purification procedures are likely to be highly complex.

Bibliography

AIBA, S., HUMPHREY, A. E. & MILLIS, N. F., (1965), *Biochemical Engineering*, Academic Press, New York.

AIBA, S. & NAGATANI, M., (1971), 'Separation of cells from culture media', *Adv. Biochem. Eng.*, **1**, 31–54.

BIROLAUD, P., BLACHERE, H., GUILBOT, A. & MOSSE, J., (1968), *Progres en Chimie Agricole et Alimentaire*, Hermann, Paris.

BUNGAY, H. R. & BUNGAY, M. L., (1968), 'Microbial interactions in continuous culture', *Adv. Appl. Microbiol.*, **10**, 269–288.

CASIDA, L., (1968), *Industrial Microbiology*, John Wiley and Sons, New York.

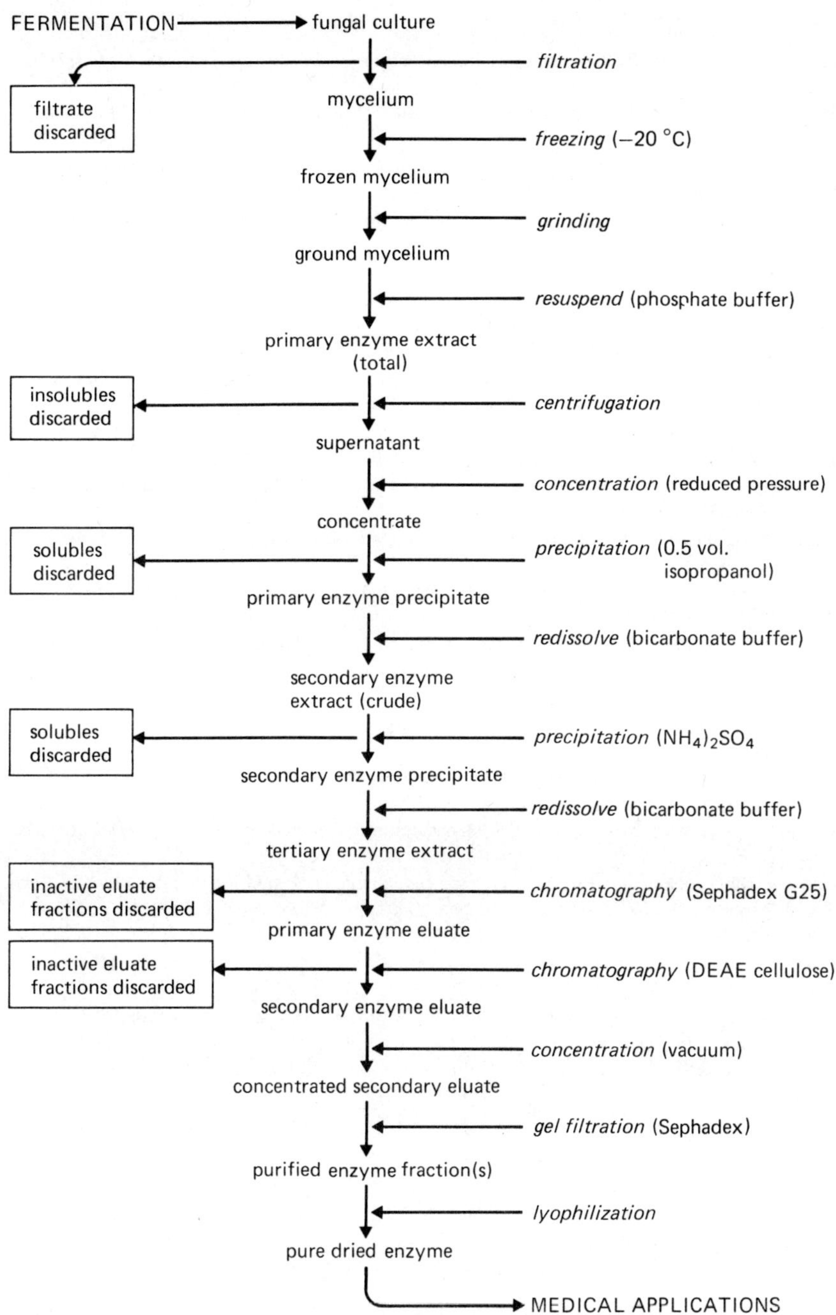

Figure 3.24 Simplified scheme for preparation of a purified enzyme (endocellular)

DEINDOERFER, F. H., (1960), 'Fermentation kinetics and model processes', *Adv. Appl. Microbiol.*, **2**, 321–333.

EDWARDS, V. H., (1969), 'The recovery and purification of biochemicals', *Adv. Appl. Microbiol.*, **11**, 159–207.

FORREST, W. W. & WALKER, D. J., (1971), 'The generation and utilization of energy during growth', *Adv. Appl. Microbiol.*, **13**, 213–269.

GEHRIG, R. F. & KNIGHT, S. G., (1958), 'Formation of ketones from fatty acids by spores of *Penicillium roqueforti*,' *Nature*, **182**, 1237.

GERHARDT, P. & BARTLETT, M. C., (1959), 'Continuous industrial fermentation', *Adv. Appl. Microbiol.*, **1**, 215–255.

GREENSHIELDS, R. N. & SMITH, E. L., (1974), 'The Tubular Reactor in Fermentation,' *Process Biochem.*, **9**, April, 11–17.

HALL, M. J., DICKINSON, S. D., PRITCHARD, R. & EVANS, J. I., (1973), 'Foams and foam control in fermentation processes', *Progr. Ind. Microbiol.*, **12**, 169–234.

HEROLD, M. & NECASEK, J., (1959), 'Protected fermentation', *Adv. Appl. Microbiol.*, **1**, 1–18.

HUMPHREY, A. E., (1960), 'Air sterilization', *Adv. Appl. Microbiol.*, **2**, 302–310.

KLEIN, F. *et al.*, (1971), 'Growth of pathogenic virus in a large scale tissue culture system', *Appl. Microbiol.*, **21**, 265.

KNIGHT, S. G., (1966), 'Transformation: a unique enzymic activity of mould spores and mycelium, *Ann. N.Y. Acad. Sci.*, **139**, 8–15.

MARSHECH, W. S., (1971), 'Current trends in the microbiological transformation of steroids', *Progr. Ind. Microbiol.*, **10**, 49–104.

MAXON, W. D., (1960), 'Continuous fermentation', *Adv. Appl. Microbiol.*, **2**, 335–349.

NYIRI, L. K., (1972), 'Applications of computers in biochemical engineering', *Adv. Biochem. Engin.*, **2**, 49–96.

ROBINSON, R. F. & DAVIDSON, R. S., (1959), 'The large scale growth of higher fungi', *Adv. Appl. Microbiol.*, **1**, 261–272.

RYU, D. D. Y. & LEE, B. K., (1975), 'An example of process optimization of enzymic transformation of steroids,' *Process Biochem.*, **10**. Jan/Feb, 15–19.

SCHULTZ, J. S. & GERHARDT, P., (1969), 'Dialysis culture of microorganisms,' *Bacteriol. Revs.*, **33**, 1–47.

SIMON, P. & MEUNIER, R., (1970), *Microbiologie Industrielle et Genie Biochemique*, Masson, Paris.

S. C. I. Monograph No 12, (1961), *Continuous Culture of Microorganisms*, Society of Chemical Industry.

STOUDT, T. H., (1960), 'Microbiological transformations of steroids,' *Adv. Appl. Microbiol.*, **2**, 183–218.

VEZINA, C., SEHGAL, S. N. & SINGH, K., (1968), 'Transformation of organic compounds by fungal spores,' *Adv. Appl. Microbiol.*, **10**, 221–265.

WANG, D. I. C. & SINSKEY, A. J., (1970), 'Collection of microbial cells,' *Adv. Appl. Microbiol.*, **12**, 121–150.

4 Microbial Proteins

The technology discussed in this chapter is that of using rapidly multiplying microorganisms, capable of utilizing readily available organic materials as their carbon and energy substrate, to convert cheap inorganic nitrogen compounds into nutritionally valuable cellular proteins to replace the more expensive plant and animal proteins used in animal and human diets.

In the late 1950's, predictions were made, on the best evidence available, of an impending global food protein shortage. Various childhood diseases associated at least in part with protein malnutrition, including kwashiorkor and marasmus, had been identified in underdeveloped countries. In 1955, the Protein Advisory Group was created to help the World Health Organization (WHO) to advise FAO and UNICEF on the safety and suitability for human consumption of new protein foods. The Group's attention was first directed to the development of protein-rich food mixtures based on dried skim milk, soya flour or locally available sources of vegetable protein. Later, interest developed in the potential of microbially-produced protein as an animal feed supplement and, possibly, as a human dietary constituent.

The emphasis has now shifted somewhat. Childhood malnutrition diseases, while still widespread, are no longer generally accepted by nutritionists as purely protein deficiencies, but are viewed as 'energy-protein malnutritions' with many other factors of child care playing a major part. The global shortage now anticipated is not merely of proteins, but rather of food of all kinds.

Nonetheless, microbial proteins ('single cell protein', SCP) may be expected to play a significant part in efforts to make good the protein component of world food shortages, initially as a cheap and rapidly manufactured protein feed for farm livestock such as cattle, pigs and poultry. The growth rate of microorganisms varies according to species, but is commonly high (Table 4.1). Under very favourable conditions, the generation time of certain bacteria may be as short as $0.3–2.0$ h and that of yeasts, $1–3$ h. Algae can double their biomass in $2–6$ h and filamentous fungi in $4–12$ h.

Microbial cells generally have a high protein content ($7–12$ g protein N per 100 g dry weight) and microorganisms synthesize proteins much more rapidly than higher plants and animals (Table 4.2).

Many microorganisms can utilize cheap, inorganic nitrogen compounds

Table 4.1 Growth rates of some microorganisms

Microorganism	*Growth rate* *(days^{-1})*	*Culture medium*	*Protein (% dry weight)*
Bacteria			
Escherichia coli	59·7	Lactose broth	82
Pseudomonas fluorescens	29·9	Glucose broth	
Azotobacter chroococcum	13·8	Urea + glucose	
	3·0	Mineral salts + sucrose	
Yeasts			
Hansenula anomala	13·8	Glucose + yeast extract	53
Protozoa			
Tetrahymena geleii	4·15	Yeast autolysate	
Algae			
Chlorella pyrenoidosa	1·96	$CO_2 + NO_3^-$	
Scenedesmus quadricauda	2·03	$CO_2 + NO_3^-$	
Euglena gracilis	1·38	$CO_2 + NH_4^+$	

such as ammonium salts for their growth, together with cheap carbon and energy substrates, producing a biomass of useful protein content. Something very similar takes place in the rumen of ruminant animals, where the natural micro-flora and -fauna can utilize cellulose and urea, producing assimilable or digestible substances which are ultimately transformed by the animal into meat and milk.

Table 4.2 Daily production rates for various forms of food protein

Organism (1000 kg)	*Protein produced (kg/day)*	*Daily yield (%)*
Beef cattle	1	0.1
Soya	10	1
Yeast	10^5	10^4
Bacteria	10^{11}	10^{10}

In microbial protein production, the desired cell composition is high protein, low carbohydrate, low nucleic acid, low lipid. The product has to compete economically with plant-derived proteins and should therefore contain favourable balances of lysine, methionine and tryptophan, which plant proteins often lack.

4.1   Substrates utilized

Carbon dioxide serves as the carbon source for algae and also for the hydrogen bacteria. Some bacteria and also certain fungi (*Graphium, Trichoderma*) can utilize methane or methanol. Industrial or agricultural wastes and by-products can also be utilized, for example, molasses, milk whey or sulphite liquor (a by-product of paper pulp manufacture); suitable strains are necessary, the most used being *Saccharomyces cerevisiae, Candida utilis* and *Saccharomyces fragilis*. A waste product from some food industries is an aqueous solution containing starch. Many yeasts are unable to break down starch and those that can are not particularly suitable for food protein production. However, a mixed culture of *Endomycopsis fibuliger* and *Candida utilis* can be grown on starchy waste liquor, the former producing an amylase, the latter rapidly utilizing the sugars liberated. Starting with an inoculum

Table 4.3   Genera of yeasts able to utilize aliphatic hydrocarbons for growth

n-alkanes (paraffins)			*1-alkenes (olefins)*
Candida	Mycotorula	Torulopsis	Candida
Cryptococcus	Pichia	Trichosporon	Debaryomyces
Endomyces	Rhodotorula	Saccharomyces	Hansenula
Hansenula			Rhodotorula

Table 4.4   Genera of filamentous fungi able to utilize aliphatic hydrocarbons for growth

n-alkanes (paraffins)		*1-alkenes (olefins)*
Absidia	Gliocladium	Aspergillus
Acremonium	Graphium	
Aspergillus	Helicostylum	Cephalosporium
Botrytis	Helminthosporium	
Cephalosporium	Monilia	Cunninghamella
Chaetomium	Mucor	
Chloridium	Oidiodendron	Fusarium
Cladosporium	Paecilomyces	
Colletotrichum	Penicillium	Helminthosporium
Cunninghamella	Rhizopus	
Dematium	Scolecobasidium	Spicaria
Epicoccum	Spicaria	
Fusarium	Syncephalastrum	
	Trichoderma	

containing equal numbers of the two species, and using urea as the nitrogen source, the material harvested contains ten times as much *Candida utilis* as *Endomycopsis fibuliger*. This process is being tried as a pilot operation in Sweden where it is known as the Symba process. Hydrocarbons are utilized as substrates by many yeasts and fungi (Tables 4.3 and 4.4).

4.2   Culture conditions

Ammonium salts or nitrates are most commonly used to provide the nitrogen source. The pH of the medium is generally made acid for yeasts (4·5–5·5) and neutral for bacteria (6·0–7·5); for the blue-green bacterium *Spirulina maxima* the medium is alkaline (9–11). The optimum temperature varies, according to the strain employed, from 28°C–40°C. It should be remembered that heat is generated during microbial growth; cooling of the fermenters can be a real problem in tropical countries (as, for example, in yeast molasses fermentation in Cuba) for there may be little difference between the temperature of the available cooling water and that of the fermenting liquid. Aseptic fermentation conditions can sometimes be dispensed with, particularly with yeasts growing in acid media: the BP (Lavéra) process for producing yeast from gas oil does not employ sterile culture conditions. Again, when algae are cultivated in open-air lagoons, bacterial and protozoal contamination always occurs.

Some special problems occur when hydrocarbons are used as growth substrate. Their low solubility has already been mentioned: Table 4.5 is repeated from Chapter 2, for convenience.

Table 4.5   Solubility of *n*-alkanes in water at 25°C

Alkane	Molar concentration in saturated solution
Hexane	$1·1 \times 10^{-4}$
Octane	$5·8 \times 10^{-6}$
Decane	$3·1 \times 10^{-7}$
Dodecane	$1·7 \times 10^{-8}$
Tetradecane	$9·8 \times 10^{-10}$

The fermenter therefore contains four phases, the aqueous medium, the suspended hydrocarbon, the growing microbial cells and the aeration gas. Since hydrocarbons contain no oxygen, the oxygen component of the biomass has to be furnished entirely by aeration (a 25% component in bacteria, 30% in yeasts). The oxygen requirement for hydrocarbon fermentations is therefore high, 2·5–3·5 times (or 4–5 times if the hydrocarbon used is methane) that needed for glucose. Naturally, the evolution of heat is correspondingly greater (Table 4.6).

Table 4.6 Influence of substrate and cell yield on oxygen requirement and heat production

| | | | | Heat released | |
Microorganism	Substrate	Cell yield (g/l)	O_2 required (g/100 g cells)	(kcal/ 100 g cells)	(kJ/100 g cells)
Yeasts	carbohydrates	0·5	67	380	1591
Yeasts	n-alkanes	1·0	197	799	3345
Bacteria	n-alkanes	1·0	172	780	3266

The chemical composition of the cell crop is influenced by the nature of the medium and by other cultural conditions. For example, the proportion of protein to lipid is affected by the C:N ratio in the medium; when the N content of the medium is low enough to become the limiting factor for growth, lipids (or, in some cases, poly β-hydroxy-butyrate) accumulate in the cells (Table 4.7).

Table 4.7 Lipid content of microbial cells grown in media containing limiting quantities of nitrogen

Microorganism:	Rhodotorula	Nocardia	Chlorella
% Lipid content:	60	70	80

Furthermore, the growth rate of a culture affects not only the chemical composition of the cells (DNA, RNA, protein) but also their size (Table 4.8).

Table 4.8 Effect of growth rate on cell size and chemical composition in *Salmonella typhimurium*

Growth Rate	Number of cells (10^{12}/g dry weight)	Cell composition (mg/g dry weight)		
		DNA	Protein	RNA
2·4	1·3	30	670	310
1·2	3·1	35	740	220
0·6	4·8	37	780	180
0·2	6·3	40	830	120

4.3   Nutritional value

Table 4.9 summarizes available data on the general composition of the main groups of microorganisms usable as food; it should be noted that 10–15% of the total N occurs in the purine or pyrimidine bases of the nucleic acids.

Table 4.9   Average composition of cells of the major groups of microorganisms (% dry weight)

	Filamentous fungi	Algae	Yeasts	Bacteria
Nitrogen	5–8	7·5–10	7·5–8·5	11·5–12·5
Lipid	2–8	7–20	2–6	1·5–3
Ash	9–14	8–10	5–9·5	3–7
Nucleic acids	—	3–8	6–12	8–16

Amino N represents 70–80% of the total N of microbial cells. Microorganisms synthesize the essential amino acids required for human growth and nutrition and Table 4.10 compares the essential amino acid content of various microorganisms with that of wheat protein and egg albumin. The amino acids shown in Table 4.10 are those required by human adults; for young children, histidine and arginine are also essential.

Table 4.10   Essential amino acid content of wheat, egg albumin and some food microorganisms (g per 16 g N)

Amino acids	Wheat	Egg white	Spirulina maxima	Saccharomyces cerevisiae	Candida lipolytica (gas oil)	Pseudomonas (methanol)	Alcaligenes eutrophus	Penicillium notatum
Lysine	2·8	6·5	4·6	7·7	7·8	5·3	8·6	3·9
Threonine	2·9	5·1	4·6	4·8	5·4	4·5	4·5	—
Methionine	1·5	3·2	1·4	1·7	1·6	1·8	2·7	1
Cystine	2·5	2·4	0·4	—	0·9	0·3	—	—
Tryptophan	1·1	1·6	1·4	1·0	1·3	—	1·1	1·25
Isoleucine	3·3	6·7	6·0	4·6	5·3	3·9	4·6	3·2
Leucine	6·7	8·9	8·0	7·0	7·8	7·0	8·5	5·5
Valine	4·4	7·3	6·5	5·3	5·8	5·9	7·1	3·9
Phenylalanine	4·5	5·8	5·0	4·1	4·8	4·2	4·0	2·8

Egg albumin is a well balanced source of the essential amino acids for human nutrition and the microbial proteins compare well, except that they contain less of the sulphur-containing amino acids. In general, bacteria tend to produce somewhat more methionine than do yeasts. However, yeasts tend to contain more lysine (an essential amino acid in which wheat protein is significantly low).

Vitamin content

The vitamins produced by microorganisms are mainly of the B group: yeasts are particularly rich in these vitamins. Vitamin B_{12} is an exception, commercially useful quantities being mainly produced by bacteria. Algae contain β-carotene (a precursor of vitamin A) and tocopherols (vitamin E) in addition to B-group vitamins. The filamentous fungi appear to have a lower vitamin content; *Morchella hortensis* is included in Table 4.11 as an example of a higher fungus.

Table 4.11 Vitamin content of some food microorganisms (mg/100 g dry weight)

Vitamin	Morchella hortensis	Candida utilis	Saccharomyces cerevisiae	Methylomonas methanica
Thiamine	0·52	0·53	5–36	1·81
Riboflavin	1·31	4·50	3·6–4·2	4·82
Niacin	12·4	41·73	32–100	15·9
Pyridoxine	2·62	3·34	2·5–10	14·3
Pantothenic acid	12·6	3·72	10	2·42
Choline	4·61	—	—	968·0
Folic acid	1·09	2·15	1·5–8	—
Inositol	1·78	—	—	—
Biotin	0·015	0·23	0·5–1·8	—
Vitamin B_{12}	0	0	0	0·96
p-aminobenzoic acid	—	1·7	0·9–10	—

Determination of protein value

In human nutrition, the essential amino acids (those which man is unable to synthesize in sufficient quantity) are obtained by the ingestion of protein foods (Table 4.12). The value of a particular protein as a source of dietary nitrogen is largely determined by its amino acid composition.

If one of these amino acids is lacking, or is present in insufficient quantity in the diet, protein synthesis is impaired. For example, rice is deficient in lysine, and children on an exclusively rice diet fail to grow normally. Quite a number of protein-containing foods are deficient in one or more amino acids (Table 4.13).

Table 4.12 Daily requirements (g) of essential amino acids for the human adult

Essential amino acids	FAO recommendation	Minimum
Phenylalanine	2·2	1·1
Methionine	2·2	1·1
Leucine	2·2	1·1
Valine	1·6	0·8
Lysine	1·6	0·8
Isoleucine	1·4	0·7
Threonine	1·0	0·5
Tryptophan	0·5	0·25
Total	12·7	6·35

Table 4.13 Nutritive value of different foodstuffs

Foodstuff	Limiting factor(s)	% Deficiency
Whole egg	(none)	0
Beef	cystine + methionine	29
Cows' milk	cystine + methionine	32
Rice, wheat, maize	lysine	61–72
Bakers' yeast	cystine + methionine	55

Moreover, the cytoplasmic proteins contained in intact microbial cells may not be readily available for the human or animal digestive processes, and the actual availability can be found only by feeding experiments. The initial tests are customarily carried out in rats which are fed the microbial protein under examination as the sole source of dietary nitrogen. The total quantity of microbial protein ingested by the animals is measured and its nitrogen content I determined by analysis; over the same period, faeces and urine are collected from the animals and their nitrogen content F and U found by analysis. Three values can then be calculated:

Digestibility D, which is the percentage of the total nitrogen consumed which is absorbed from the alimentary tract:

$$D = 100 \times \frac{I - F}{I}$$

Biological Value BV, which is that percentage of the total nitrogen assimilated which is retained by the body, taking into account the simultaneous loss of endogenous nitrogen through excretion in the urine:

$$BV = 100 \times \frac{I - (F + U)}{(I - F)}$$

Protein Efficiency PE, which is the proportion of nitrogen retained when the protein under test is fed compared with that retained when a reference protein, such as egg albumin, is fed.

Many trials of the digestibility of microbial proteins have been carried out: they tend to show that algae are of somewhat low digestibility (50–75%) while yeasts may reach 95%. Values of about 80% are quoted for *Escherichia coli* and *Acinetobacter calcoaceticus* and 90% for *Alicaligenes eutrophus*. However, pretreatment of bacterial cells before incorporation into a ration may be required to improve digestibility: for example, disruption of the cells of *Bacillus megaterium* increase digestibility from 56% to 67%. Rumen protozoa, which are of great importance in the nitrogen nutrition of cattle and sheep, have a digestibility of 86–96%. Table 4.14 summarizes some data concerning yeasts.

Table 4.14   Nutritive value of various yeast proteins

Yeast species	Species	Digestibility (%)	Biological value (%)
Saccharomyces cerevisiae	rat	81	59
Candida utilis	rat	85–88	32–48
Saccharomyces cerevisiae	man	79–90	52–87

Bacterial proteins are richer in sulphur-containing amino acids than yeasts, and consequently have a greater biological value. By supplementing yeast protein with methionine, its biological value can be increased (Table 4.15).

Table 4.15   Nutritive value of dried *Candida utilis* for rats

Ration	Digestibility (%)	Biological value (%)	Protein efficiency
Candida only	85	48	0.9
Candida + 0.5% DL-methionine	90	90	2.3

It is only possible to use yeasts which are rich in lysine and threonine (see Table 4.10) to balance a diet based on cereals (such as wheat, rice or tapioca) which have a lower content of these amino acids. Table 4.16 summarizes the results of a feeding experiment illustrating this point.

Table 4.16 Effect of adding dried *Candida utilis* to a rice
flour diet for rats (animals 46 g body weight at the
start of the trial)

Candida utilis (g/100 g diet)	% *protein in diet*	*Weight gain* (g)	*Protein efficiency*
0	6·5	30	1·87
2	6·9	36	2·19
4	7·3	68	2·90
6	7·8	95	3·13
8	8·2	108	3·29
10	9·9	122	3·03
12	11·0	137	2·93
14	11·9	132	2·75

Human feeding trials

When it is hoped ultimately to use a particular microbial protein as human
food, long term feeding trials on laboratory animals followed by autopsy and
painstaking pathological examination are necessary to reveal potential
toxicity or other harmful effects. Human feeding trials are then indicated,
cautious and on a small scale at first, with the initial emphasis on testing
acceptability; once acceptability has been established, the product is more
fully tested to reveal any liability to induce allergies or gastro-intestinal
disturbances, and also to ensure that there is no specific toxicity for man which
was not revealed by the animal trials.

There have been a few experiments in which humans have been fed algae,
mostly for periods of only 3 to 6 days. The voltunteers tolerated 100 g per day
of a mixture of *Chlorella* and *Scenedesmus* but when the intake was increased
to 200 g, gastro-intestinal upsets (nausea, vomiting, flatulence and diarrhoea)
occurred. Alcohol extracts of *Chlorella pyrenoidosa* were well tolerated by
some subjects, but not extracts of *Scenedesmus obliquus*.

Feeding studies with bacterial proteins have been rather unsuccessful in
man. Yeasts, however, are much better tolerated. Human volunteers have been
able to consume 135 g of dried yeast per day over a nine day period without
gastro-intestinal disturbances. Nonetheless, it is generally preferable to subject
microbial cells to various forms of pretreatment before employing them as
food materials.

Treatment of cells for use as protein source

Factors which may impair the usefulness of microbial proteins include their
possession of a non-digestible cell wall, high nucleic acid content, unaccept-
able colouration (especially with algae) and disagreeable flavours (algae and

yeasts). It is usually necessary to kill the cells before consumption because live cells may survive or even multiply in the intestine, giving rise to fermentation, liberating toxic amines or utilizing B group vitamins at the host's expense.

Treatment is thus directed towards killing the cells, partial or complete removal of the cell wall and reduction of nucleic acid content. Various chemical procedures have been tried. For example, urea has been used to break hydrogen bonds, forming low molecular weight urea-protein complexes, the urea being subsequently removed by dialysis. Again, pretreatment with sodium carbonate increases extraction yield by 30% in the case of *Chlorella* or 50% in *Torula* but causes breakdown of lysine, cystine and tryptophan, thus diminishing protein efficiency. Hydrolysis using strong acids (HCl) lowers the content of some amino acids and destroys tryptophan. Removal of the large amount of sodium chloride produced during neutralization is an additional problem and the procedure is chiefly applicable when the aim is to produce purified samples of a given amino acid.

Enzyme treatment has also been tried. Treatment of *Scenedesmus obliquus* cells with cellulase gives only slight improvement in digestibility, due to the presence of non-cellulose constituents in the cell walls of *Chlorophyceae*. Snail (*Helix pomatia*) gastric secretion has no effect on intact cells of *Chlorella pyrenoidosa*, though it does hydrolyze cell wall material removed from the cells; it breaks down yeast cell walls but even so is too expensive for commercial use. Cheap enzyme techniques for degrading microbial cell walls will probably depend on finding microbial sources of the enzymes. However, autolysis (in which thick suspensions lose their cell walls due to digestion by the cells' own endoenzymes) has long been used in the commercial production of yeast extract. Another form of autolysis (sometimes loosely termed plasmolysis) can occur when cells are suspended in strong (25%) sodium chloride solution and is also used in making yeast extract. Both procedures are lengthy, requiring 12–24 hours at 45–50°C, pH 6·5 under aseptic conditions and the yields are none too high (50–60%).

Cells can also be mechanically disrupted by milling, grinding with glass beads or abrasives, sonic vibration with abrasives or ultrasonic vibration. Another method uses the shearing effect produced when cell suspensions frozen at $-25°C$ are forced through narrow dies at pressures of up to $4000\,kg/cm^2$. Such procedures are costly: they require special equipment, the expenditure of energy and, since they are usually carried out on cell slurries of 10% dry weight, the subsequent removal of a large amount of water. However, since they enable removal of the non-assimilable cell wall N, the final protein product has a higher digestibility (Table 4.17).

Cheaper and simpler methods are sometimes effective. Autoclaving of algae and desiccation of bacteria and yeasts damage the cell walls to a varying extent and may increase nutritive value. In a recent procedure, yeast cells are mixed

Table 4.17 % digestibility of whole cells compared with that of cell contents released by disruption

Species	Whole cells	Disintegrated cells
Bacillus megaterium	76	94
Candida utilis	65	95

with edible oil and heated to 170°C; the treated cells are then separated from the mixture giving a product of 1–5% water content, 7–15% lipids and an agreeable taste and odour.

Reduction of nucleic acid content

Nucleic acids tend to be present in high concentration in cells having a high growth rate. Microbial cells may contain 8–25 g nucleic acids per 100 g protein; in liver the concentration is 4 g or less. In the gut, nucleic acids are first attacked by pancreatic nuclease; the nucleotides so formed are hydrolyzed to nucleosides by nucleotidases secreted by the intestinal mucous membrane (Figure 4.1).

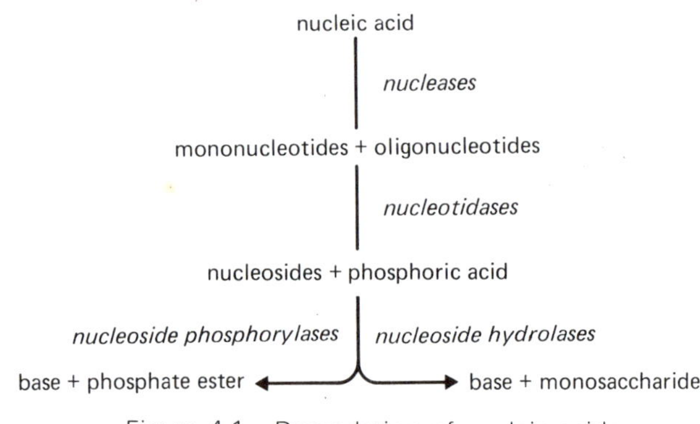

Figure 4.1   Degradation of nucleic acids

Figure 4.2 shows the path by which the purine bases of nucleic acids are degraded prior to excretion. In man, the end point is uric acid, for the human species has, during evolution, lost the ability to synthesize uricase (urate oxidase) which in many other vertebrates transforms the sparingly soluble uric acid into highly soluble allantoin.

The main danger is that if human food contains too much nucleic acids, the excess of uric acid formed will not be excreted by the kidneys and may

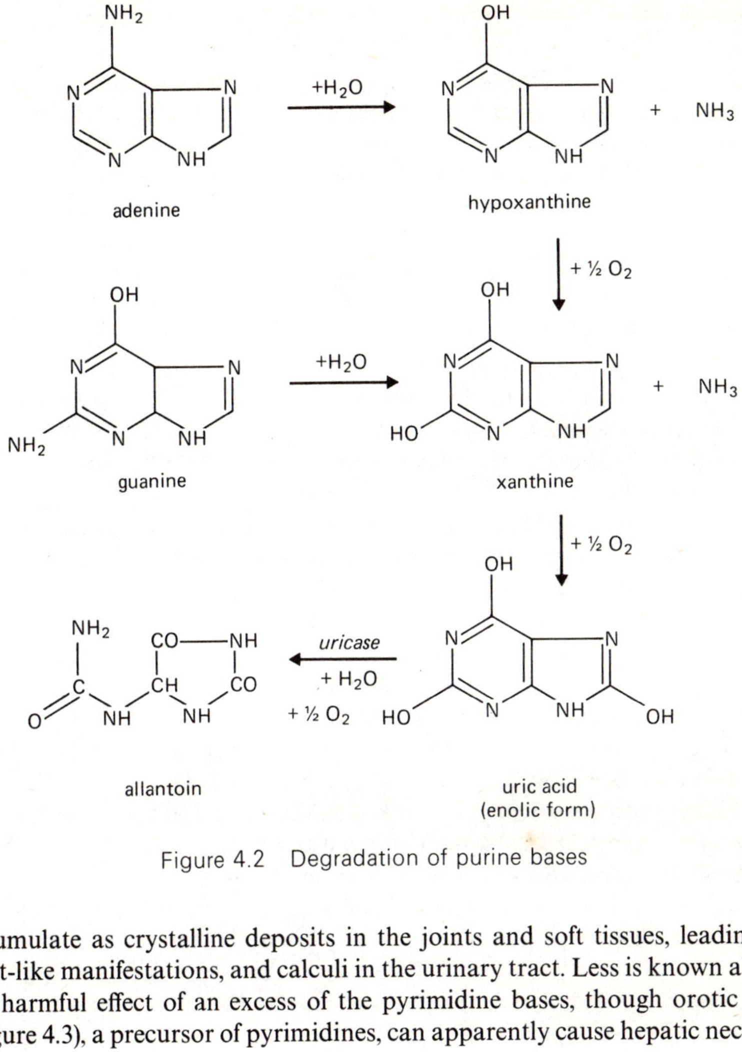

Figure 4.2   Degradation of purine bases

accumulate as crystalline deposits in the joints and soft tissues, leading to
gout-like manifestations, and calculi in the urinary tract. Less is known about
the harmful effect of an excess of the pyrimidine bases, though orotic acid
(Figure 4.3), a precursor of pyrimidines, can apparently cause hepatic necrosis
in rats.

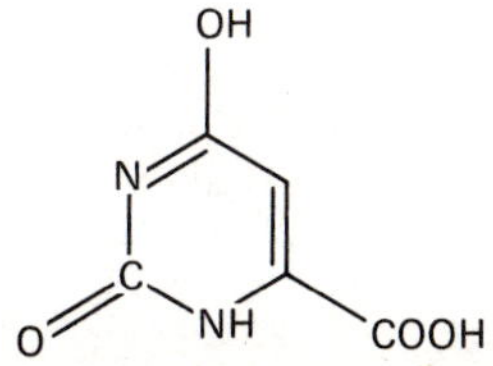

Figure 4.3   Orotic acid

Many trials on human volunteers have been made. In one, the volunteers consumed 130 g yeast daily for a week after which the concentration of uric acid in the plasma reached 4·8–8·3 mg per 100 g (normal plasma levels for uric acid are generally quoted as 2–7 mg/100 ml for males, slightly less for females); the uric acid content of the urine doubled also. In another trial, 20 g of microbial protein containing 8·5% nucleic acid was well tolerated. In general, it would appear that a daily intake of 2 g nucleic acid is harmless for a healthy person.

It has been suggested that the human species' lack of urate oxidase (and the resulting higher levels of plasmic uric acid) might in some way be connected with the superiority of human intelligence over that of other vertebrates. There is some evidence of correlation between uric acid plasma levels and some types of brain function. For example, rats fed yeast ribonucleic acid were found to perform certain tests better than control rats on a normal diet. There have also been claims of successful treatment of premature senility in man by oral or intravenous administration of sufficient ribonucleic acid to produce plasma levels of 7–9 mg/100 g. Unfortunately, the brain superiority hypothesis fails to account for the fact that, like man, other higher primates, Dalmatian dogs, birds and reptiles (save for tortoises) lack uricase.

Fanciful theories apart, the risk of injury by uric acid deposition is real and much effort has been concentrated on reducing the nucleic acid content of microbial proteins. The cellular RNA content can be lowered by markedly reducing the growth rate (see Table 4.8) but this measure sacrifices the main advantage of using microorganisms, namely speed of production. Other methods have been sought.

Alkaline hydrolysis destroys cellular RNA but also diminishes the nutritive value of the subsequently extracted proteins or amino acid mixture. Some RNA can be extracted from yeast cells with 10% sodium chloride: the most efficient method is to disintegrate a cell suspension at pH 9·5 and heat to precipitate protein with a nucleic acid content of less than 2%; sodium

Table 4.18 Effect of treatments to reduce nucleic acid content in a strain of *Candida utilis* (results given as % dry weight)

| | | Supernatant from cells treated sequentially with: | | Centrifuged deposit |
Cell constituent	*Untreated cells*	Heat	Ribonuclease	(Microbial protein)
Nucleic acids	7·5–9·0	2·0–3·0	5·5–6·0	1·0–2·0
Protein	45–55	4·0–5·5	0·3–0·5	40–50
Amino acids	—	1·0–1·1	0·2–0·4	—

chloride extraction is then applied. With *Candida utilis*, thermal shock treatment (30–80°C) followed by digestion with pancreatic ribonuclease can reduce the nucleic acid content from $7\cdot5$–$9\cdot0\%$ in the intact cells to 1–2% in the final centrifuged protein (Table 4.18).

Mammalian pancreatic ribonuclease is expensive: microbial phosphodiesterases, which can be obtained cheaply by fermentation, would be an acceptable and economic substitute.

Another technique, applicable to yeasts, is to activate the endo-enzymes by a sequence of mild thermal shocks. An initial few seconds at 68°C denatures the ribosomes and activates ribonucleases. Reduction of temperature to 45°C for 2 hours followed by 1 hour at 55°C encourages RNA hydrolysis and the diffusion of low molecular weight products out of the cells. The RNA content can be reduced from 7% to 1% (dry weight) by this method. The RNA, if extracted without damage, can be used to produce the nucleotides guanylic monophosphate (GMP) and inosinic monophosphate (IMP) which have a use as flavour-enhancing agents.

Toxicity testing

Before any new microbial protein product can be used in animal feeds, let alone as a human dietary constituent, comprehensive testing is necessary to ensure that it is devoid of toxicity or other harmful effects. There is a special problem with yeasts or other organisms grown on hydrocarbons obtained from the petro-chemical industry for such substrates may contain carcinogens including 3,4-benzpyrene, 1,2,5,6-dibenzanthracene and methylcholanthrene. The actual concentration of these substances in the harvested yeast is lower than $0\cdot0002$ ppm, one hundredth of that which may be found in smoked meats; nonetheless, the remote possibility of carcinogenesis makes long-term animal feeding trials a necessity.

Sometimes, disappointing results in feeding trials are due to factors other than toxicity and can be remedied once the cause is understood. For example, liver lesions sometimes develop in rats fed on a yeast diet and this was thought to be due to an amino acid imbalance; in fact, it arose from a deficiency in selenium and vitamin B_{12} and could be avoided when selenium was added to the yeast culture medium. Again, when beer yeast or *Candida utilis* was fed to rats over several generations, some samples appeared to yield slower growth of the test animals, cannibalism was observed and also infertility in females of the second generation; if the deficiency of sulphur-containing amino acids was rectified, results were comparable with those achieved by conventional feeds.

Feeding trials in laboratory animals are followed by comparable tests in the food animals. Thus, hens fed 20% of yeast grown on oil have been shown to lay normal quantities of eggs, and these have in turn been fed to rats without causing harm. Pigs also have been fed in this way over long periods, with no detectable effect on growth, gestation or lactation.

4.4   Different types of microbial protein

Proteins obtained from bacteria, yeasts, filamentous fungi and algae respectively will now be discussed.

Bacterial proteins

Hydrogen bacteria
The hydrogen bacteria are aerobes which obtain energy by the oxidation of H_2 and acquire carbon from CO_2. The genus *Hydrogenomonas* is not now recognized, and hydrogen oxidizing bacteria have been assigned to other genera as follows: *Pseudomonas facilis, P. saccharophila, P. ruhlandii, P. flava, P. palleronii, Alcaligenes eutrophus* and *A. paradoxus*. The former *Hydrogenomonas carboxydovorans* is probably a *Pseudomonas*. (Some anaerobic bacteria utilize the same substrates, but are obligate phototrophs with slow growth rates that preclude their industrial use.) The name *Pseudomonas facilis* signifies the ease with which this and other species can be isolated. Enrichment cultures can be obtained in a mineral salts medium containing, per litre of distilled water:

NH_4Cl	1 g	$MgSO_4.7H_2O$	0·2 g
$Na_2HPO_4.12H_2O$	9 g	$Fe(NH_4)$ citrate	5 mg
KH_2PO_4	1·5 g	$CaCl_2 2H_2O$	10 mg
$NaHCO_3$	0·5 g		

100 ml bottles containing 15 ml of this medium are inoculated with 1 g of soil and incubated at 30°C, with agitation, in an atmosphere of 70% H_2, 20% O_2 and 10% CO_2. After several days, subcultures are similarly made from those bottles showing growth. Final purification is by streaking onto the same medium solidified with the addition of 2% of washed agar. Gram negative, motile, peritrichate rods are regularly recoverable. Besides the recognized species, there are various unnamed strains (H16, H1, H20, etc.).

Hydrogen-oxidizing species are generally mesophilic, but a few strains have an optimum growth temperature of 50°C. Their nutritional requirements are simple and ammonium salts, nitrates or urea serve as a nitrogen source: K^+, Mg^{2+}, Ca^{2+}, Fe^{3+}, Ni^{2+}, PO_4^{3-} and SO_4^{2-} are also required; the optimum pH is 6·5–7·5. However they also grow well on media containing organic acids and amino acids and are therefore only facultative chemo-autotrophs. They lack glucose permease; fructose is the only sugar which serves as a carbon and energy source. Glucose is formed within the cell by isomerization of fructose and is degraded via the Entner-Doudoroff pathway. Growth can also occur under anaerobic conditions in the presence of nitrate, which acts as a hydrogen acceptor. The presence of H_2 suppresses synthesis of the enzymes of the Entner-Doudoroff pathway and generally inhibits utilization of organic substrates; thus, in the presence of H_2 and CO_2 metabolism is always autotrophic.

Grown autotrophically at 33–35°C, the hydrogen bacteria display a generation time of 3·5–7 hours and yields of 28 g/l have been obtained after 10 days, higher than with other chemo-autotrophic bacteria. In continuous cultures, exponential growth ceases above 3 g/l dry weight, but this cell density can be maintained over periods of months. The oxygen requirement is high, $0·66 l O_2$ per hour per g dry weight yield; thus, a culture containing 10 g/l of cells (dry weight) consumes 300 mmoles (6·6 l) O_2 per hour. Table 4.19 shows that either H_2 or O_2 supply may serve as limiting factors for growth; the higher cell yield occurs when the limiting factor is O_2.

Table 4.19 H_2 and O_2 as limiting factors for growth of a hydrogen bacterium, H.16, in a chemostat

Parameter	*Limiting factor*	
	H_2	O_2
Growth rate (h^{-1})	0·08	0·08
Cell density (mg/ml)	0·167	0·237
Poly β-hydroxybutyrate (% dry wt.)	<1	23
H_2/CO_2 ratio	9·1	4·6

A gas mixture of suitable proportions ($H_2/O_2 = 2$) can be furnished by electrolysis of the mineral salts medium within the fermenter, using platinum electrodes. Explosion risks are then obviated, for the gas mixture is consumed by the bacteria as it is produced. The reactions involved are as follows:

(1) Energy $\qquad\qquad\qquad 4H_2 + 2O_2 \rightarrow 4H_2O + 224\,kcal$

(2) CO_2 assimilation $\qquad 2H_2 + CO_2 \rightarrow (HCHO) + H_2O$

(3) Overall reaction $\qquad 6H_2 + 2O_2 + CO_2 \rightarrow (HCHO) + 5H_2O$

(4) Electrolysis of water $\quad 6H_2O \rightarrow 6H_2 + 3O_2$

During electrolysis, 1 mole of H_2 is produced by 53·6 ampere hours, i.e. 1 kWh yields $1000/53·6 = 18·65$ moles of H_2. Incorporation of 12 g C requires

6 moles of H_2 (see reaction 3); thus, each kWh used in electrolysis allows assimilation of 36 g C, corresponding to about 66 g of cell dry weight having a protein content of about 50%. The process could therefore be viewed as a conversion of electrical energy to bacterial protein.

Besides their protein, the cells contain 15 g nucleic acid and 20 g cell wall material per 100 g (dry weight). The amino acid composition of *Alcaligenes eutrophus* is given in Table 4.20: for comparison, that of casein is included and also that of a strain of *Pseudomonas saccharophila* grown in an organic medium, though this latter species can also grow in a mineral salt medium using the oxidation of H_2 as the energy source.

Table 4.20   Amino acid composition of *Alcaligenes eutrophus* and *Pseudomonas saccharophila* compared with that of casein (g amino acid per 16 g N)

Amino acid	Casein	A. eutrophus	P. saccharophila
Tryptophan	1·34	—	—
Threonine	4·30	4·52	5·37
Lysine	8·06	8·61	5·73
Methionine	3·10	2·69	2·03
Cystine	0·38	—	0·36
Isoleucine	6·59	4·58	4·14
Leucine	10·11	8·52	8·35
Phenylalanine	5·42	3·96	3·56
Tyrosine	5·86	3·26	2·32
Valine	7·44	7·13	7·55
Histidine	3·04	2·48	1·81
Arginine	4·10	8·00	5·01
Alanine	3·38	8·80	13·57
Aspartic acid	7·44	9·57	9·72
Glutamic acid	2·32	11·17	10·52
Glycine	2·00	5·47	9·65
Proline	11·82	3·46	5·59
Serine	6·69	3·47	4·64

Table 4.21   Nutritive value of *Alcaligenes eutrophus*

Protein source	Digestibility	Biological value
A. eutrophus (boiled)	93·8	77·0
A. eutrophus (ultrasonicated)	93·3	77·0
E. coli	81·0	69·0
Casein	98·9	77·1

Alcaligenes eutrophus protein is low in the sulphur containing amino acids but, like that of yeasts, is rich in lysine; its digestibility and biological value, as determined in rats, is comparable with that of casein (Table 4.21).

Many hydrogen-utilizing strains of bacteria accumulate in their cytoplasm reserve granules of poly β-hydroxybutyrate (PHB), stainable with Sudan black, and this occurs particularly when the medium is deficient in N or P or when O_2 is the limiting factor. This polymer is not assimilated by animals and appears unchanged in the faeces. Unfortunately, when maximum biomass production is achieved by making O_2 the limiting factor (Table 4.19), the PHB content of the cells approaches 25%. Results of some attempts to select PHB-less mutants are shown in Table 4.22.

Table 4.22 Poly β-hydroxybutyrate (PHB) content of a hydrogen-oxidizing bacterium, H16 (wild strain), and five mutants (limiting quantities of nitrogen)

	PHB (% dry wt.) after incubation with:			
Strain	Fructose (40 h)	Gluconate (23 h)	Acetate (23 h)	$CO_2 + H_2$ (26 h)
H16 (wild type)	65·3	27·7	37·2	35·8
H16 PHB⁻ strain 1	11·9	7·3	—	5·6
H16 PHB⁻ strain 2	—	8·6	8·2	1·5
H16b PHB⁻ strain 3	—	7·3	13·6	4·3
H16 PHB⁻ strain 4	—	0	0	0
H16 PHB⁻ strain 5	—	—	0	0

Hydrogen bacteria offer promise as a useful protein source, though economical production processes remain to be worked out. A particular attraction is that the substrates (CO_2, H_2, O_2, mineral salts) carry no possibility of contamination of the biomass with extraneous carcinogens.

There is a further point of interest. The reactions for electrolytic culture of hydrogen-oxidizers show that an overall excess of O_2 is produced: reaction 3 (see above) uses two molecules of O_2 but reaction 4 produces 3 molecules. Photosynthesis, too, leads to the production of oxygen:

$$H_2O + CO_2 \rightarrow (CHOH) + O_2$$

There have been studies by the American NASA on the feasibility of such microbial systems for O_2 regeneration in space capsules during manned interplanetary flight (Table 4.23). In general it would seem that the CO_2 produced by one astronaut's respiration could be absorbed—and sufficient O_2 regenerated—by a continuous culture (turbidostat) system containing 20–30 l of a bacterial suspension with 10 g/l dry weight of cells. It is believed that some such arrangement has been provided for Soviet astronauts during prolonged earth-orbitings.

Table 4.23 Requirement for O_2 regeneration in space capsules (for one individual for 270 days)

System	Weight (kg)	Power (kW)
Chemical regeneration		
LiOH	125	1·40
NaOH	155	7·68
CO_2–H_2	34	0·36
Biological regeneration		
Algae (artificial light)	116	10·40
Algae (sunlight)	103	1·70
Bacteria (electrolysis)	55	0·75

Bacteria utilizing methane or methanol

Methane is cheap, abundant and without direct toxicity. It is a constituent of North Sea gas and it is also produced during the anaerobic digestion of sewage. Methane is the most highly reduced form of carbon; when used for bacterial growth, high cell yields relative to the amount of gas consumed are possible, and any excess of unconsumed gas is easily recovered since it is only slightly soluble in water.

Methane utilizing bacteria are found in water (including sea water), soils and muds; they are often present in the upper, aerobic layers of marshy land where they oxidize methane produced by bacteria in the deeper layers. Two genera of bacteria which utilize one-carbon organic compounds only are currently recognized, and both are Gram negative, strictly aerobic chemoorganotrophs. *Methylomonas* consists of motile rods and *Methylococcus* of non-motile cocci. Other genera have been described (*Methylosinus, Methylocystis, Methylobacter*) but their taxonomic status is uncertain. In addition, members of such genera as *Bacillus, Pseudomonas, Vibrio* and *Brevibacterium* have been described as utilizing both one-carbon and multicarbon compounds.

The best known species of true methylobacterium is *Methylomonas methanica;* it produces pink colonies, forms an extracellular slime and is catalase and oxidase positive. Isolation is difficult. Enrichment cultures are made in a liquid mineral salts medium (composition given in Table 4.24), followed by plating onto an agar version of the same medium; the presence of organic compounds other than methane or methanol may inhibit growth. The cultures are incubated in closed vessels containing methane (10–98%), oxygen

(2–45%) and CO_2 (0–20%); recommended proportions vary considerably. Protozoa may grow in the liquid enrichment media and non-methane utilizing bacteria may develop on the plates as a result of impurities present in the agar.

Table 4.24 Media for isolation and continuous culture of methane utilizing bacteria

Constituents	For isolation (g/l)	For continuous culture (g/l)
KH_2PO_4	1·60	0·67
Na_2HPO_4	1·16	0·22
$NaNO_3$	1·18	9·55
$MgSO_4, 7H_2O$	0·080	0·32
$FeSO_4, 7H_2O$	0·014	0·029
$Ca(NO_3)_2, 4H_2O$	0·075	0·18
$CuSO_4, 5H_2O$	$4·0 \times 10^{-3}$	$9·1 \times 10^{-3}$
$ZnSO_4, 7H_2O$	$3·4 \times 10^{-4}$	$1·1 \times 10^{-3}$
$MnSO_4, H_2O$	$3·0 \times 10^{-4}$	$1·5 \times 10^{-3}$
$Na_2MoO_4, 2H_2O$	$2·4 \times 10^{-4}$	$6·4 \times 10^{-4}$
$CoCl_2, 6H_2O$	—	$4·5 \times 10^{-5}$
H_2SO_4 (36 N)	—	2·7 ml

Methanol-mineral salts media may also be used but, in fact, for some methylobacteria even 0·1% of methanol is toxic. Nitrates or ammonium salts can serve as a N source and the pH range is 5·8–7·4, the optimum being about 6·8. Optimum growth temperatures range from 15°C to 35°C according to strain. High-yielding thermophilic strains also occur, growing at 45–55°C, particularly in the methylococci.

Oxidation of methane occurs thus:

$$CH_4 \xrightarrow[-26·12]{} CH_3OH \xrightarrow[-44·81]{} HCHO \xrightarrow[-57·15]{} HCOOH \xrightarrow[-58·25]{} CO_2$$

$$(\text{kcal mole}^{-1} \text{ at pH 7})$$

$$-109·36 \qquad -187·62 \qquad -239·29 \qquad -243·89$$

$$(\text{kJ mole}^{-1} \text{ at pH7})$$

The evidence for this series of reactions is as follows: washed suspensions of methylobacteria are capable of oxidizing the intermediate compounds; addition of certain inhibitors causes accumulation of the intermediates, e.g. 0·02 M sodium sulphite produces accumulation of formaldehyde and 0·003 M iodoacetate produces methanol; cell extracts can be shown to contain methanol, formaldehyde and formic acid dehydrogenases.

Electron microscopy has shown that methane and methanol utilizing bacteria possess complex membraneous structures within their cytoplasm resembling, in this respect, many autotrophic bacteria. It is possible to distinguish two types of membranes. Those possessing type I membranes (arranged in bundles throughout the cell) utilize formaldehyde by the allulose phosphate pathway (Figure 4.4); those with type II membranes (arranged in pairs throughout the cell or at its periphery) utilize the serine pathway (Figure 4.5). Both paths have the same energy efficiency and yields are similar. The allulose pathway occurs in *Methylomonas methanica* and *Methylococcus capsulatus* and the serine pathway in *Methylomonas methanooxidans* and *Pseudomonas* AM-1.

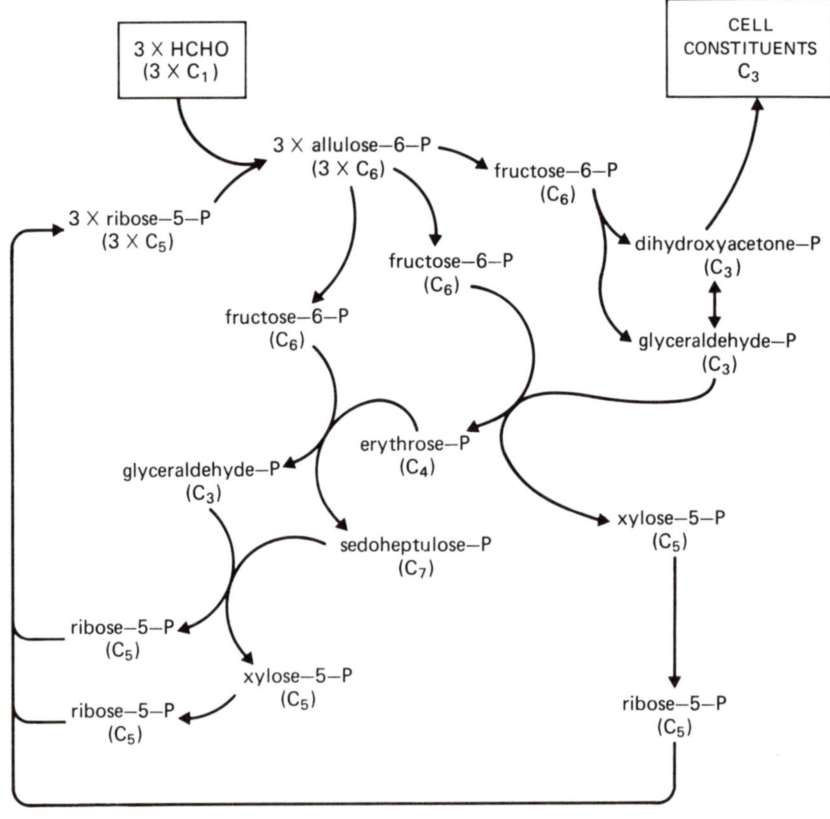

Figure 4.4   The allulose phosphate pathway

On an industrial production scale, certain difficulties arise from the use of methane as a substrate: these are the risk of explosion, and the low solubility of methane which gives rise to problems of mass transfer across the cell

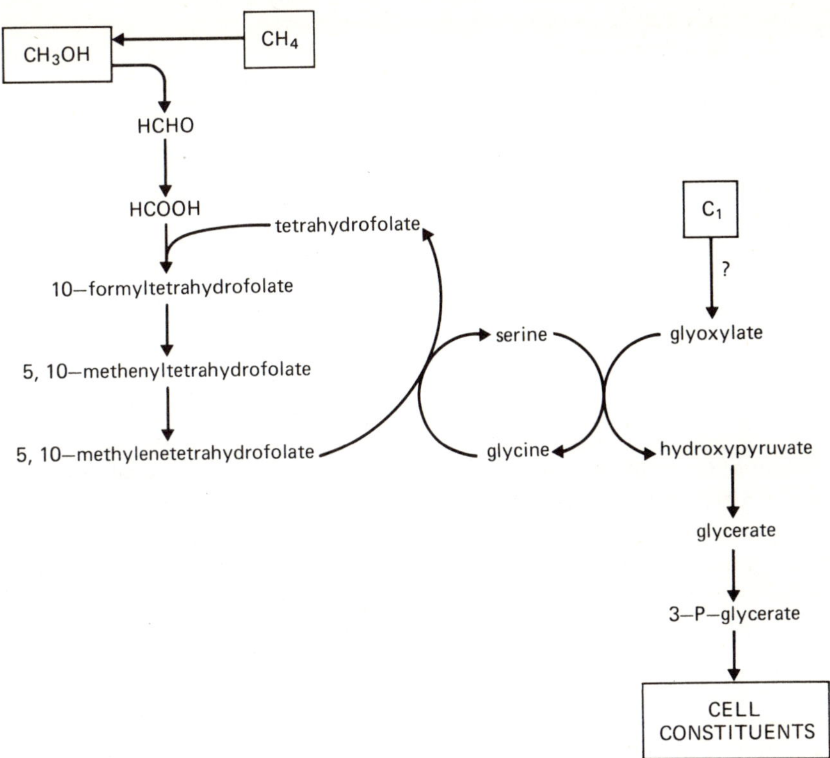

Figure 4.5   The serine pathway

membrane. Other difficulties arise from the protracted generation time of the methylobacteria (3–16 h) and the large quantity of heat released during substrate utilization (Table 4.25).

When pure cultures are used, yields are generally low (1 g dry weight per litre, or less) but many experiments have shown that mixed cultures of two different strains (always Gram negative rods) can give much higher yields, of the order of 12 g/l. In one continuous culture experiment, the generation time reached 2·29 hours at maximum growth rate, corresponding to a dilution rate of 0·303 h^{-1}. The mean cell yield (dry weight) per gram of methane consumed, Y(CH$_4$), was 0·616 g and the dry weight per gram oxygen Y(O$_2$), was 0·215 g. Maximum production of 2·39 g cells/l/h was obtained when the dilution rate was 0·187 h^{-1}, the bacterial density being 12·8 g/l. The limiting factor is the rate of mass transfer of oxygen (Table 4.25).

The difficulty of maintaining the required mixed populations of methane utilizing bacteria in equilibrium over a long period of culture is a disadvantage.

Table 4.25 Continuous fermentation using a mixed culture of methylobacteria

Dilution rate	Bacterial density			Gass consumption (ml/min/litre)	
(h^{-1})	(g/l)	$Y(CH_4)$	$Y(O_2)$	CH_4	O_2
0·220	6·83	0·618	0·221	81·0	99·7
0·154	11·74	0·592	0·197	97·7	132·3
0·100	1·61	0·641	0·214	26·7	8·53

However workers at Shell Research have developed techniques for the controlled growth of reconstituted defined mixed cultures on methane and methanol on an industrial scale (Harrison, *et al.*, 1976). They suggest that the use of chemostat techniques for selecting cultures from soil and water samples to grow on novel substrates, usually gives rise to stable communities rather than pure cultures. The members of such a community appear to be dependent on each other and are usually not able to utilize the substrate provided. Thus the mixed culture used at Shell Research to produce biomass from methane contains an as yet unidentified methane utilizing bacterium which produces methanol and other metabolites. The methanol acts as a substrate for the growth of a species of *Hyphomicrobium* whereas the other metabolites are removed by two other species of bacteria (*Flabobacterium* sp. and *Acinetobacter* sp.). The dependence of three of these organisms on the methane-utilizing species is obvious, but it is not immediately apparent why the methane bacterium itself should be dependent on them. Most probably it requires the removal of its own metabolites for optimum growth.

Methanol is almost as cheap as methane, and can readily be produced from methane present in natural gas by various limited combustion processes. When methanol is used as the growth substrate, the mass transfer problem associated with methane disappears, since methanol is completely miscible with water. Moreover, the O_2 requirement is somewhat reduced, as methanol contains an atom of oxygen; also, much less heat is evolved and cooling costs are less. Table 4.26 illustrates some of the advantages of methanol.

In addition to bacteria, a few yeasts have been found which grow aerobically on methanol, though at a lower pH (3·5–4·0). They include *Kloeckera* species, *Candida silvicola* and *Torulopsis glabrata*.

The oxygen requirement for growth on methanol, though lower than for methane, is still considerable and oxygen transfer tends to be the limiting factor. One solution could be the use of oxygen-enriched air or even pure oxygen as the aeration gas; the greater expense of this could be offset by

Table 4.26 Oxygen uptake and heat release during continuous culture
of bacteria on various carbon sources

Carbon source	Cell yield on substrate (g cell/g substrate)	Cell yield on oxygen (g cell/g oxygen)	Oxygen demand (mmole/l/hr)	Heat of fermentation (kcal/l/h)	(kJ/l/h)
Glucose	0·5	2·1	53	6·4	26·8
Methanol	0·5	0·7	150	18·4	77·0
Methane	0·6	0·2	560	67·0	280·5
n-alkanes	1·0	0·5	210	25·0	104·7

reduced power costs for agitation and gas pumping. However, with one strain
at least (*Pseudomonas* AM-1), high levels of dissolved O_2 (above 420 mm Hg)
actually depress the yield.

The protein content of methanol grown cells is 35–70 g/100 g (dry weight).
Amino acid profiles for a *Pseudomonas* (TM 20) grown on methanol and a
mixed culture grown on methane are given in Table 4.27. They do not differ
greatly from each other or from those of other bacterial proteins.

Table 4.27 Amino acid profiles for bacterial proteins
produced from methanol and methane
(grams amino acid/16 g N)

Amino acid	*Pseudomonas* TM 20 (methanol)	*Mixed culture* M45 (methane)
Cystine	0·32	0·35
Histidine	1·73	1·6
Isoleucine	3·90	5·0
Leucine	6·96	7·9
Lysine	5·30	4·3
Methionine	1·81	3·0
Phenylalanine	4·18	5·2
Threonine	4·52	4·9
Tryptophan	—	2·7
Tyrosine	2·91	3·8
Valine	5·85	6·2

A process developed by ICI utilizes a Gram negative rod given the name of
Methylophilus methylotrophus, grown continuously on a sterile medium
containing methanol (produced from natural gas), inorganic salts and ammonia. The pressure cycle fermenter employed has several unique features: the
large quantity of air supplied is used to produce rapid internal circulation and

thus no stirring mechanism is needed; also, the height of the fermenter creates a high pressure region at the bottom, where sterile air is introduced, allowing rapid dissolving of oxygen, and a low pressure region at the top which causes release of carbon dioxide. A suspension containing about 3% dry weight of bacterial cells is continuously removed; the cells are concentrated by a flocculation and flotation process (recovered water being recycled), and dried (Figure 4.6).

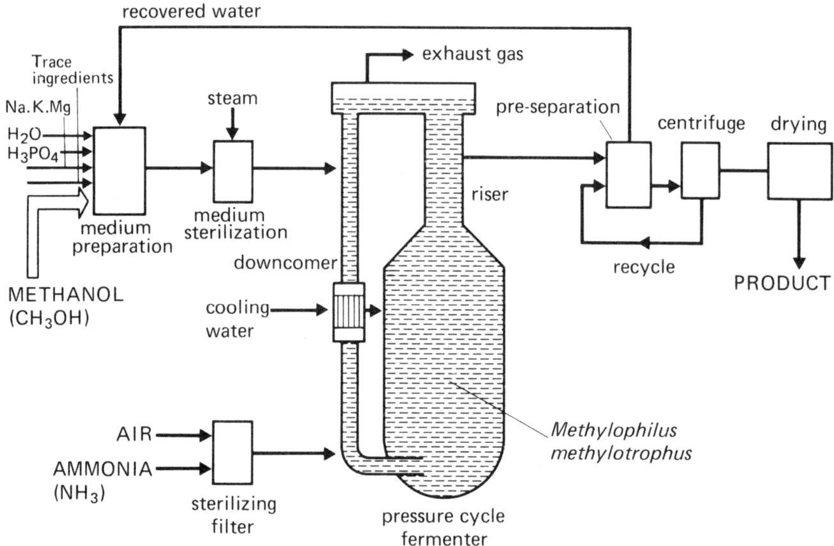

Figure 4.6   Production of bacterial protein from methanol

The product ('Pruteen') contains 72% protein and 8·6% total lipids, with an amino acid profile high in lysine and methionine and comparable with that of fish meal (Table 4.28). Feeding trials were carried out on rats, poultry, pigs, calves, lambs, mink and fish and, in some of these species, included multi-generation studies on physical and reproductive performance. The nucleic acid content of the bacterial product is excreted as uric acid in poultry and as allantoin in other species. A commercial plant capable of producing 100,000 tons annually is to be constructed at Billingham, UK.

Bacteria utilizing n-paraffins

Hydrocarbon-utilizing bacteria have been isolated from soil samples by research workers of the ESSO group, using media containing hydrocarbons as the sole carbon source (Table 4.29).

Table 4.28 Amino acid profiles of *Methylophilus* protein

Amino Acid	Grams amino acid/ 16 grams Nitrogen Methylophilus protein	Grams amino acid/100 g product		
		Methylophilus 72%	fish meal 66%	soya meal 47%
Lysine	6·8	4·9	4·9	2·9
Methionine	2·5	1·8	1·9	0·6
Cystine	0·7	0·5	0·7	0·6
Alanine	7·0	5·0	4·2	2·1
Arginine	5·0	3·6	3·7	3·5
Aspartic acid	9·0	6·5	6·2	5·4
Glutamic acid	10·5	7·6	8·9	8·7
Glycine	5·3	3·8	3·9	2·1
Histidine	2·0	1·4	1·5	1·4
Isoleucine	4·8	3·4	3·2	2·1
Leucine	7·6	5·2	5·0	3·6
Phenylalanine	3·6	2·6	2·9	2·4
Proline	3·4	2·4	2·9	2·2
Serine	3·4	2·4	3·0	2·6
Threonine	4·6	3·3	3·0	2·5
Tryptophan	1·0	0·7	0·9	0·6
Tyrosine	3·3	2·3	2·3	1·9
Valine	5·6	4·1	3·7	2·1

Table 4.29 Some hydrocarbon-utilizing bacteria

Strain	Protein (% of dry weight)	Index of essential amino acids
Brevibacterium insectiphilium ATCC 15528	53·6	60
Corynebacterium sp. ATCC 15529	69·0	58
Corynebacterium paurometabolum ATCC 15530	59·6	63·3
Pseudomonas ligustri ATCC 15522	60·9	46·8
Pseudomonas pseudomallei ATCC 15523	56·8	58·2
Pseudomonas orvilla ATCC 15524	62·0	63·0
Alcaligenes sp. ATCC 15525	70·0	74·1
Cellulomonas galba ATCC 15526	51·4	65·9

In Taiwan, many similar strains have been isolated by comparable methods. The most promising is a strain of *Pseudomonas*, 5401, which utilizes *n*-paraffins from crude oil or various refined fractions, ammonium salts acting as the nitrogen source; spontaneous emulsification of the hydrocarbons in the culture medium occurs after about 4 hours fermentation. After 24 hours at 36–38°C growth density reaches 16 g/l (dry weight) as shown in Figure 4.7.

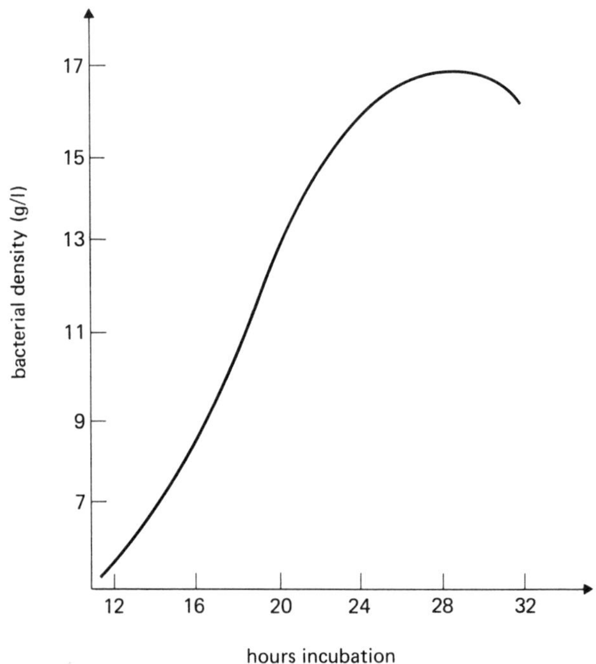

Figure 4.7   Growth curve of a *Pseudomonas* (strain 5401) in a liquid medium containing *n*-alkanes as carbon source

In continuous culture of *Pseudomonas* 5401, a dilution rate of $0.12\,h^{-1}$ maintains growth in the exponential phase, with a steady cell concentration of 10 g/l. The cell crop is harvested by centrifugation, treated with solvent to remove residual hydrocarbon and dried and ground. The protein content can be as high as 73·62 g/100 g dry weight with an essential amino acid composition better than that of yeast, in particular having a higher content of lysine and methionine (Table 4.30). The dried cells also contain various B group vitamins (Table 4.31).

Table 4.30 Essential amino acid composition of various microbial proteins produced using *n*-alkanes (g/100 g dry weight)

Amino acid	Yeast	Pseudomonas 5401	Brevibacterium insectiphilium	Corynebacterium paurometabolum	Alcaligenes sp.	Cellulomonas galba
Arginine	2·37	4·98	2·72	3·0	4·1	3·0
Glycine	2·45	3·75	2·0	2·4	3·0	2·6
Isoleucine	2·35	3·34	2·3	3·4	1·8	2·5
Leucine	3·60	6·57	3·5	3·7	4·8	3·0
Methionine	—	1·27	0·72	0·75	1·2	0·75
Phenylalanine	2·17	3·29	1·5	1·7	1·8	1·7
Valine	2·60	3·48	2·5	2·6	3·8	2·7
Tryptophan	0·27	1·16	—	—	1·0	1·2
Lysine	3·89	4·29	2·4	2·3	5·0	0·9

Table 4.31 Vitamin content of *Pseudomonas* 5401 grown on hydrocarbons

Vitamin	Content (µg/g dry weight)
Thiamine (chlorhydrate)	2·88
Riboflavin	10·50
Niacin	180
Calcium pantothenate	7·0
Pyridoxine	1·6

Yeasts

Attempts to use yeasts as food date from the shortages of the first World War and were renewed during the second. The value of dried yeast as a supplement for both animal and human diets is well established. World production is of the order of 350,000 metric tons per year, comprising 150,000 tons of bakers' yeast with the remainder being used for animal feeds.

Yeasts are able to synthesize amino acids from inorganic nitrogen and sulphur compounds (ammonium salts and sulphates). Their carbon and energy source can be provided by agricultural or industrial by-products such as molasses, starchy materials (which usually have to be hydrolyzed), milk whey, fruit pulps and the sulphite liquor obtained during the manufacture of paper pulp. Breweries produce large quantities of yeast (*Saccharomyces cerevisiae*) as a by-product. Various yeast species are used, depending on the nature of the carbon substrate, but the gross composition of the dried cells is very similar (Table 4.32).

Table 4.32   Gross composition of various food yeasts (g/100 g dry weight)

Constituent	S. cerevisiae molasses	Candida utilis sulphite liquor	S. fragilis milk whey	S. cerevisiae beer
Protein	50	55	54	45
Lipids	6	5	1	6
Moisture	5	6	7	6
Ash	7	8	9	8
Sodium	0·3	0·001	—	0·2

Candida utilis has been much employed because of its rapid growth rate, its ability to utilize pentose sugars (present in sulphite wastes), its non-exacting nutritional needs and because large-celled, readily separable variants exist. A process developed in Sweden (the Symba process) used the α- and β-amylase-producing *Endomycopsis fibuliger* to hydrolyze starch to sugars which were then utilized by *Candida utilis*. *Saccharomyces diastaticus* forms an extra-cellular glucoamylase and can utilize partly-hydrolyzed starch, though it appears to have been little used.

Yeasts grown on molasses seem to have the highest content of essential amino acids (Table 4.33) though the sulphur-containing amino acids are low by comparison with casein or egg albumin. Yeasts grown anaerobically have a reduced content of essential amino acids, and it should be remembered that beer yeast undergoes a major part of its development under anaerobic conditions.

The yeast cells are harvested by centrifugation and thoroughly washed to remove traces of the medium which may otherwise impart undesirable flavour or colour; brewers' yeast has a particular tendency to retain traces of hop bitters. The washing water may be partly recycled for economic reasons. Final harvesting is by rotary vacuum filter, yielding a cake containing some 20–40% dry matter; this is dried to give a product of 6–10% water content. Figure 4.8 and Table 4.34 summarize the treatment.

Table 4.33 Essential amino acid composition of various food yeasts (g/100 g total protein)

Amino acid	*S. cerevisiae* molasses	*Candida utilis* sulphite liquor	*S. fragilis* milk whey	*S. cerevisiae* beer
Lysine	8·2	6·7	8·8	7·3
Valine	5·5	6·3	6·6	5·2
Leucine	7·9	7·0	9·9	6·3
Isoleucine	5·5	5·3	5·5	5·7
Threonine	4·8	5·5	5·5	4·8
Methionine	2·5	1·2	1·5	1·2
Phenylalanine	4·5	4·3	3·9	4·4
Tryptophan	1·2	1·2	1·5	1·1
Histidine	4·0	1·9	2·5	1·5
Arginine	5·0	5·4	4·9	4·7

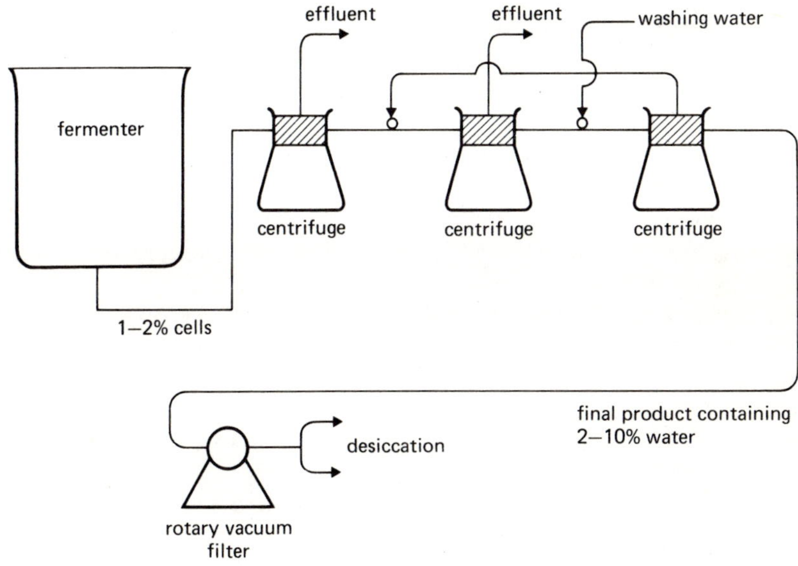

Figure 4.8 Stages in the post-fermentation treatment of food yeast

Despite the use of by-products as substrate, the production cost of food yeasts is quite high. In the search for yet cheaper substrates, attention has turned to those paraffins (*n*-alkanes) which are obtained as by-products during the distillation of petroleum.

Table 4.34   Production of food yeasts

Yeast content in fermenter (g/l)	Yeast content after centrifuging (% dry weight)	Washing water (vol/vol)	Number of washings	Type of filter used
30	15–24	1·0	1	Press
20	20	0·4	1	Rotary vacuum
10	14–18	3·0	3	Rotary vacuum

Production of yeasts from hydrocarbons

Many studies are currently being carried out in different countries. In France, for example, there has been collaboration between the French BP company (Société Française des Pétroles BP) and the Bacterial Chemistry Laboratory of the National Centre for Scientific Research in Marseilles, leading to industrial processes for producing *Candida lipolytica* (=*Saccharomycopsis lipolytica*) and *C. tropicalis* from petroleum hydrocarbons.

Hydrocarbon utilizing yeasts can readily be isolated from oil-polluted soils, or even ordinary garden or agricultural soils, using mineral salts agar media containing hexadecane as the carbon source, and 0·002% chloramphenicol to inhibit bacterial growth.

Production media for these yeasts contain inorganic N, P and S sources; ferrous and magnesium ions appear to increase the cell yield, as does the addition of yeast extract to the medium. The pH is controlled by adding ammonia, which supplements and, indeed, may provide the inorganic nitrogen source.

Published results are few but Table 4.35 summarizes one study in which it was found that the highest yield occurred when the medium contained 2% hexadecane and 0·5% ammonium nitrate.

Table 4.35   Effect of hexadecane and ammonium nitrate concentration on cell yield (ml of cell volume produced per 10 ml medium)

NH_4NO_3 %	n-hexadecane			
	0·5%	1%	2%	4%
0·25	0·20	0·31	0·46	0·50
0·50	0·20	0·36	0·68	0·67
1·0	0·20	0·36	0·48	0·64
2·0	0·18	0·32	0·46	0·62
4·0	0·16	1·30	0·40	0·61

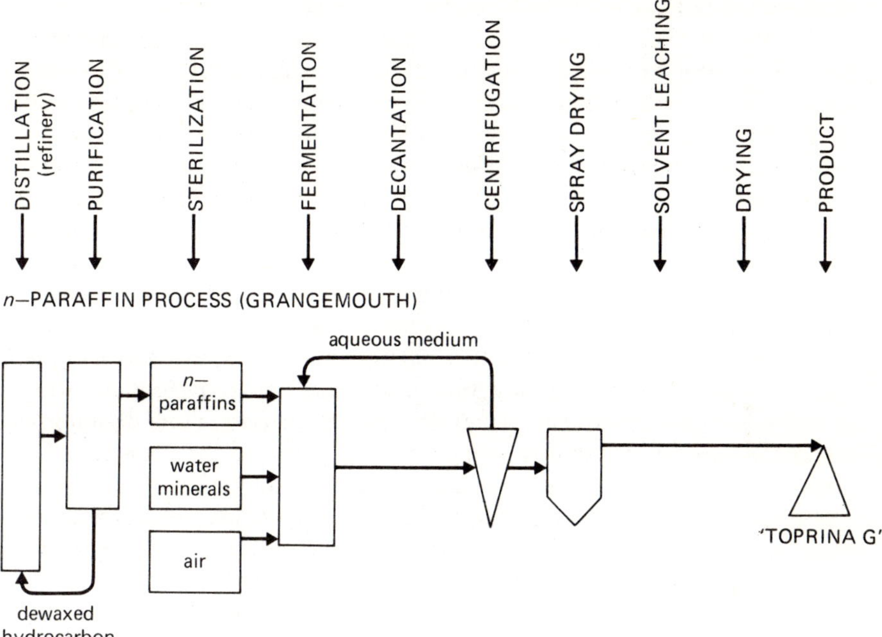

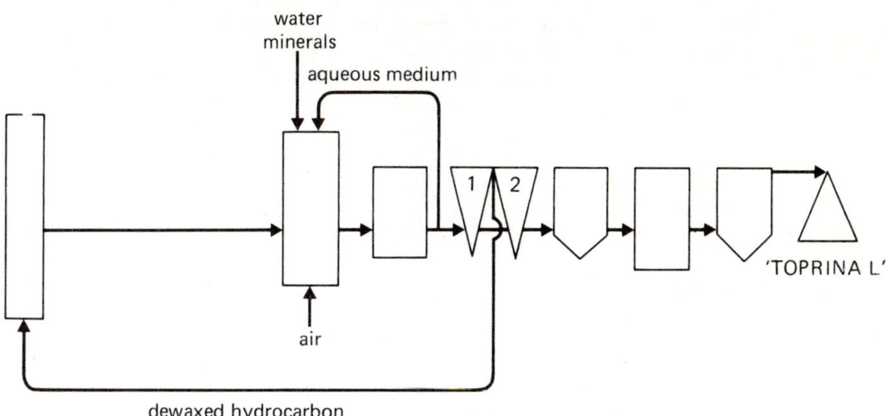

Figure 4.9   Summaries of two BP processes for production of yeast protein feeds from oil

Two processes, covered by BP patents, for the production of yeast from oil are depicted in Figure 4.9. Both employ ammonia as the nitrogen source. In the Scottish (Grangemouth) process, n-paraffins (C_{10}–C_{18}), separated from gas oil by molecular seiving, are used as the culture substrate, the residue being returned to the refinery. The fermenter is a stirred tank operating continuously by simple displacement, and all feed streams to it are sterilized. Hydrocarbons are completely utilized during fermentation. The culture passes to a yeast centrifuge and is separated into a yeast cream, which is spray dried, and a clear aqueous phase which contains some unused mineral salts and is recycled to the fermenter.

At the Lavéra plant, Bouches-du-Rhône, gas oil is injected direct into an air-lift fermenter which is open to the atmosphere. Neither the air nor the medium constituents are sterilized and culture purity is maintained by control of pH, temperature (30°C) and dilution rate. The yeast selectively uses the waxy paraffins of higher molecular weight and the culture passes to a decanter in which the aqueous medium is separated for recycling to the fermenter. The remainder, composed of yeast cells, gas oil and some of the aqueous phase, is centrifuged to yield a yeast cream and a dewaxed oil fraction which is returned to the refinery with a much improved pour point. The yeast is spray dried, but contains some residual oil: this oil, together with a large proportion of the yeast cellular lipids, is extracted by counter current leaching with solvent. The solvent is distilled for reuse and yeast lipids can be recovered as a by-product.

Table 4.36 Nutrient and power requirements for the production of 100,000 tons of yeast by the Grangemouth and Lavéra processes

Item	*Grangemouth* (*n*-paraffins)	*Lavéra* (gas oil)
Nutrients (*tons*)		
PO_4	5200	8,600
K	1800	3,300
Mg	200	280
Mn	50	40
Zn	120	230
Fe	40	20
NH_3	14,000	16,000
Power		
Air compressor (MW)	34	28
Steam (kg/h)	40,000	113,000
Fuel (kcal/h)	70×10^6	82×10^6
Cooling (kcal/h)	$c\ 110 \times 10^6$	$c\ 110 \times 10^6$

Some production details for the two processes are compared in Table 4.36 and Figure 4.9.

The product of BP Proteins Ltd. is called Toprina. Both the Grangemouth and Lavéra plants are development plants and their annual production capacity is 4000 and 20,000 tons of Toprina G and Toprina L respectively; a larger production plant of 100,000 tons Toprina G capacity is now operated in Sarroch, Sardinia by Italproteine (BP/ANIC).

These protein yeast products have been extensively tested for acute, subacute and chronic toxicity and teratogenicity (in rats), mutagenicity and carcinogenicity (in rats and mice), cutaneous sensitization (rats, guinea pigs, pigs, monkeys) and in multiple generation studies on reproduction and fertility (rats, Japanese quail) with regularly negative results. The boiling point of the carcinogenic polycyclic hydrocarbons is around 500°C: the distillation ranges of the substrates used (Grangemouth, 175–300°C; Lavéra, 300–380°C) makes it unlikely that oil-derived carcinogens could be present in the product.

However, it is known that oil-grown yeasts may contain traces (up to 0·4%) of hydrocarbons. The further treatment of these yeasts will be determined by the nature of the starting material on which they were grown. If pure n-alkane substrates are used there is no need, from the toxicological point of view, for further treatment since neither the feedstock nor the residual hydrocarbons in the yeast contain detectable levels of known carcinogens using analytical methods sensitive to 1 μg per kilogram. If it is required, for reasons other than that of toxicological acceptability, to reduce the level of residual hydrocarbons in yeasts grown on pure n-alkanes this may be effected in one of several ways. For example, the cells can be transferred to a secondary fermenter and subjected to a maturation process in the presence of air, during which they metabolize most of the adsorbed hydrocarbon. Further treatment with an assimilable, surfactant sucrose ester detaches the remaining iso- and cyclo-paraffins (Figure 4.10)

In the case of a gas oil substrate, much hydrocarbon which cannot be assimilated by the yeast becomes attached to the cells, and it is necessary to apply a solvent extraction treatment to reduce the residual hydrocarbons to an acceptable level ($<0·08\%$). After centrifuging or decanting to remove most of the aqueous culture phase, the creamy mixture of cells, medium and oil can be transferred to an extraction tank and agitated at 50–80°C with a polar solvent (isopropanol, ethanol or acetone). Hydrocarbons and yeast lipids enter the solvent phase, which is redistilled to recover the solvent and also the lipid byproduct. The cell fraction can be further extracted with a polar/non-polar solvent mixture (e.g. 20% ethanol/80% hexane). Figure 4.11 gives an outline of this form of extraction process.

In general, 1 kg of paraffins gives a yield of about 2 kg (dry weight) of yeast cells which, in turn, contain about 1 kg protein. The overall composition of the

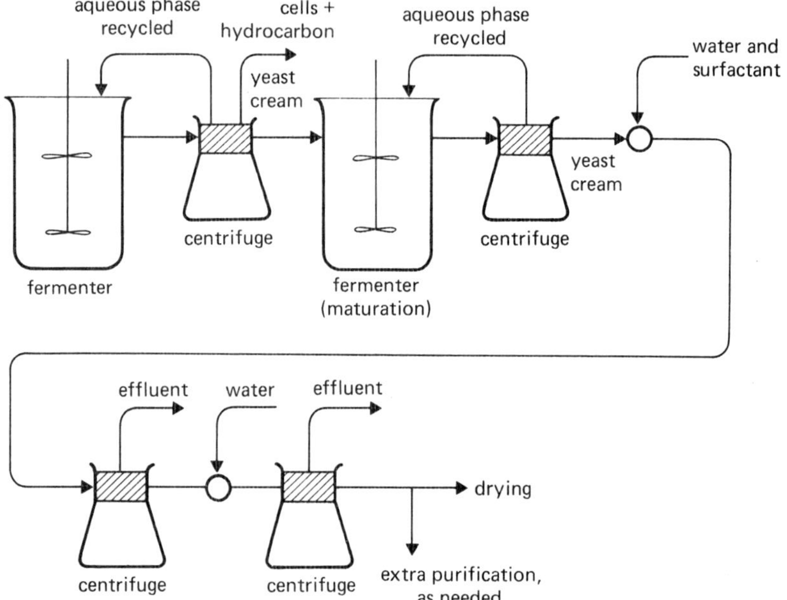

Figure 4.10   Purification of *n*-paraffin-grown yeast by maturation and washing

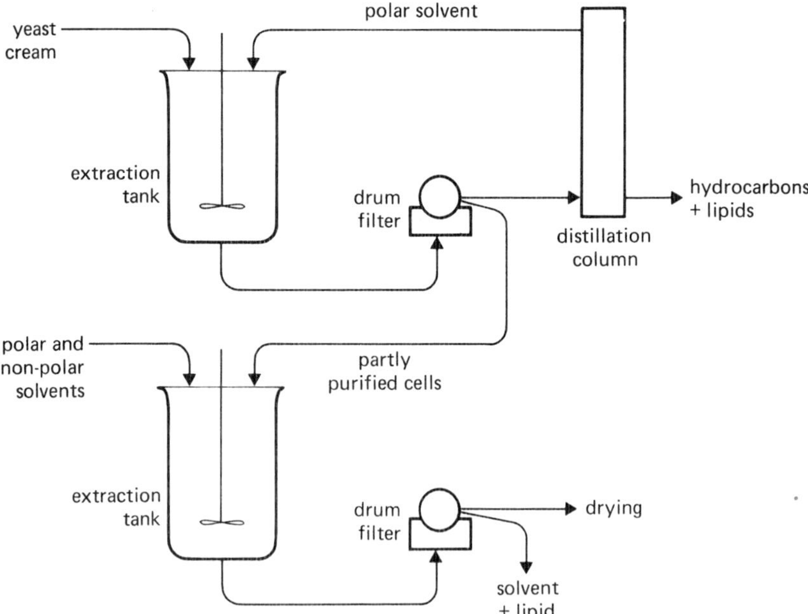

Figure 4.11   Purification of gas oil-grown yeast by solvent extraction

Table 4.37 Typical analysis of yeast
produced from oil

Item	Grangemouth	Lavéra
moisture % weight	5	5
Dry basis		
Nitrogen % weight	9·8	11·2
Crude protein % weight	61·0	70·0
Lipid % weight	10·0	1·5
Ash % weight	7·0	7·9
Phosphorus % weight	1·8	1·9
Digestibility %	80	80

BP Grangemouth and Lavéra products is given in Table 4.37 and the amino
acid composition in Table 4.38.

The content of lysine is high in oil-produced yeast proteins but that of the
sulphur-containing amino acids low. However, methionine can be added and,
as previously shown (Table 4.15), increases digestibility, biological value and

Table 4.38 Amino acid composition of yeast
produced from oil (g per 16 g N)

Amino acid	Grangemouth	Lavéra
Isoleucine	4·5	5·3
Leucine	7·0	7·8
Lysine	7·0	7·8
Phenylalanine	4·4	4·8
Tyrosine	3·5	4·0
Cystine	1·1	0·9
Methionine	1·8	1·6
Threonine	4·9	5·4
Tryptophan	1·4	1·3
Valine	5·4	5·8
Alanine	7·4	5·8
Arginine	4·8	5·0
Aspartic acid	9·2	10·0
Glutamic acid	11·3	12·1
Glycine	4·8	4·5
Histidine	2·0	2·1
Proline	4·4	3·7
Serine	4·8	5·1
Glucosamine	1·8	2·0

protein efficiency. Long term animal feeding trials of hydrocarboyeast protein products have been made in various countries including the UK, France, Holland and the USSR, with generally satisfactory results. Table 4.39 gives some data obtained in pigs.

Table 4.39 Growth of pigs fed a diet containing 18·2% crude protein over a 17 week period

Value measured	Control fish meal	7·5% BP concentrate + methionine	15% BP concentrate + methionine
Mean daily gain (g)	646	641	657
Consumption index	3·07	2·98	2·94

Tables 4.40 and 4.41 summarize the results of some trials concerning egg production in laying hens (23–78 weeks). The crude protein content of the laying ration was 16·5%. The yeast constituted the sole high protein source in the experimental ration (0–78 weeks) replacing all of the fish, sesame and soyabean meal in the control ration.

Table 4.40 Egg production during the whole laying period (till an age of 78 weeks)

Group	Number of eggs per 100 hen-days abs.	%	Average egg weight g	%	Kg egg per 100 hen-days abs.	%
Control	55·5	100	63·8	100	3·54	100
14% yeast*	60·7	109·4	62·5	98·0	3·80	107·3

*Toprina G

Table 4.41 Feed conversion and mortality during the whole laying period (till an age of 78 weeks)

Group	Feed conversion (kg feed/kg egg) abs.	%	Mortality (hens) abs.
Control	3·34	100	5
14% yeast*	3·27	97·9	4

*Toprina G

Filamentous fungi

During the second World War, attempts were made to use cultures of *Fusarium* and *Rhizopus* strains, grown in fermenters, as protein foods and subsequent research has taken place in the U.K. (Rowett Research Institute; Rank Research Centre). Cellulose is a cheap and abundant source of carbon and energy and can be utilized by mixed fungal cultures, particularly those containing the cellulolytic fungus *Trichoderma viride*. However, most work has centred on the amylolytic fungi. Barley, a basic pig food, contains about 10% protein and a ration containing 14 to 20% barley is used for growing pigs. Barley protein is deficient in lysine, and this is normally remedied by adding relatively expensive, imported animal or vegetable protein (fish or soya meal). An alternative possibility which has been investigated is to use barley, with some cheap supplementary nitrogenous material, as a medium for growing amylolytic filamentous fungi, effectively converting part of the barley starch to microbial protein. The grain is ground and mixed with water; urea or ammonium sulphate is added and the mash acidified to inhibit bacterial growth. The inoculum is either *Aspergillus oryzae* or *Rhizopus arrhizus*, chosen mainly because of their non-toxigenicity: the former is, in fact, used in the manufacture of Japanese soya-based condiments and the latter in the production of Indonesian miso.

Barley starch is hydrolyzed by the fungal enzymes to glucose, which is assimilated together with the nitrogen source and used for fungal protein synthesis. The starch/protein conversion rate is 23% for *Rhizopus* and 40% for *Aspergillus*. The growth rate of the fungi is influenced both by the starch content and the nature of the nitrogen source. Best results are obtained with strains of *Aspergillus oryzae*: after 34 h at 30°C a yield of 10.5 g mycelium (45% protein content) per litre of mash is obtained. Sterilization is not required at any stage, and the product is readily accepted by pigs when incorporated into their diet. Amino acid contents are given in Table 4.42.

Species of *Fusarium* produce proteins of high quality with a high methionine content, and might therefore be of special interest for human nutrition; they could also be used to enrich algal and bacterial proteins, which usually have a low methionine content.

Algae and blue-green bacteria

Both the eucaryotic algae and the procaryotic blue-green bacteria (cyanobacteria) utilize light energy, with CO_2 as their source of carbon; however, the CO_2 content of air is insufficient to give high growth yields. Yields of 30–40 g/l have been obtained with *Chlorella pyrenoidosa* grown under heterotrophic conditions in a glucose medium. Another approach is to use a mineral salts culture medium and pass CO_2 gas through it; the pH of the medium needs to

Table 4.42 Essential amino acid content of some fungal proteins

Essential amino acid	Rhizopus arrhizus	Aspergillus oryzae	Barley	Balanced pig ration
Lysine	6·2	7·1	4·1	5·5
Methionine + cystine	3·8	3·1	3·6	3·5
Threonine	4·0	4·3	3·6	2·5
Isoleucine	3·9	4·5	3·4	3·5
Leucine	5·9	7·2	6·8	3·5
Valine	4·8	5·2	5·1	2·5
Tyrosine + phenylalanine	6·5	6·2	8·4	2·5
Tryptophan	1·2	1·0	1·0	0·75
Histidine	2·4	2·1	2·1	1·0
Total essential amino acids	38·7	41·2	38·0	25·3

be alkaline in order that bicarbonates are formed, and the dissolved CO_2 tension is then independent of its partial pressure.

Under such conditions, during photosynthesis, the utilization of one molecule of CO_2 corresponds with the disappearance of one bicarbonate ion and the formation of an OH^- ion:

$$HCO_3^- + 2H_2O \rightarrow (CH_2O) + O_2 + OH^-$$

The pH rises, but provided further CO_2 is supplied, the bicarbonate is regenerated:

$$OH^- + CO_2 \rightarrow HCO_3^-$$

The efficiency of light energy conversion is low, at the most 4%, allowing algal yields of only 3 g per kWh. Sunlight is the only economically feasible light source. The algae have to be grown in shallow ponds or lagoons not more than 20–30 cm deep, since the turbidity they produce limits penetration of light. The biomass produced is small, of the order of 1 g/l (dry weight), requiring 0·4 g carbon, 0·1 g nitrogen and 0·01 g phosphorus: the latter two elements are provided as inorganic salts. It is not desirable that the growth should become too dense, for this prevents light penetration and losses due to algal respiration become too high. Some form of agitation may be necessary, otherwise some of the cells sink to the bottom where they receive little or no light and the remainder rise to the surface where they risk desiccation. The requisite temperature is 25–35°C, according to the strain employed. In open air ponds or lagoons it is impossible to maintain culture purity and protozoal populations, feeding on unicellular algae, may develop. Production of algae is feasible only in regions where the climatic conditions (temperature and sunlight) are suitable: the lagoons are open to the air to encourage heat loss and avoid the cost of artificial cooling during the hotter part of the year. Seasonal variations in temperature can be allowed for by selecting algal species with different optimal growth temperatures and which crop in succession (Table 4.43).

Table 4.43 Growth temperatures and yields of various algae

Species	Optimum growth temperature (°C)	Yield (kg protein per hectare per year)
Algae		
Chlorella pyrenoidosa	25	15,700
Chlamydomonas mundana	27	24,300
Blue-green bacteria		
Spirulina maxima	33	41,500
Anacystis nidulans	41	59,200

The biomass obtained is very diluted. In the case of unicellular algae, harvesting requires flocculation or centrifugation procedures. Sometimes, during hot weather, spontaneous flocculation occurs during the afternoon when both the temperature and the pH of the lagoon are rising (the latter because of accelerated CO_2 consumption); the supernatant nutrient solution can then be drained off and the sedimented algal suspension decanted. If a flocculating agent is used, this has to be later removed.

Filamentous forms, on the other hand, can easily be harvested by skimming when they rise as a result of oxygen bubbles produced during photosynthesis. Renewal of nutrient fluids is likewise simple: the spent medium is drained off and fresh pumped in. The harvested algae are separated from water trapped in the mat of filaments (usually by filtration) and then dried. The cheapest method is drying on open sand beds, but this may allow the growth of potentially pathogenic bacteria, necessitating sterilization of the product before its use as food. Suitable filamentous algae and blue-green bacteria include *Spirulina, Spirogyra, Vaucheria, Porphyra, Hormidium, Stigeoclonium* and *Uronema*.

Algae contain approximately 50 g protein/100 g dry weight (alfalfa hay by comparison contains 15%), but this protein is always low in the sulphur-containing amino acids. Algae are not good sources of vitamins of the B group, but contain plentiful β-carotene, vitamin K and some vitamin C. Man can apparently consume 30–40 g per day without trouble; indeed, 100 g per day was given in one trial in the USSR, and the natives of the shores and islands of Lake Chad have from time immemorial eaten sun-dried *Spirulina maxima*. The digestibility of algae is low and the taste can be unpleasant. The reason for the offensive taste would seem to be that part of the protein consists of enzymes which are bound to the lipids of the photosynthetic complex which contains the chlorophyl and carotenoid pigments; these enzymes bring about hydrolysis of the lipids, and the disagreeable flavour is caused by oxidation products of the fatty acids.

Results obtained from the feeding of laboratory animals with algae have not been altogether satisfactory, even with *Spirulina*. The low digestibility seems to be connected with the cell wall fraction, which comprises 20% of the dry weight and consists of enzyme-resistant complex celluloses and hemicelluloses. Dried cells are more digestible than raw ones, but maximum digestibility is obtained only by costly extraction of the cell protein (Table 4.44).

Table 4.44 % digestibility of algae for rats

Scenedesmus obliquus	*% digestibility for rats*
Untreated cells	11·1–33·4
Dried cells	70–75
Protein extract	85·6–87·4

The digestibility of a *Chlorella/Scenedesmus* mixture is 54% for pigs, but is higher (73%) for cattle, in which the rumen microflora and fauna carry out degradation of the cell wall component. Algae can therefore be used as a low-cost high-productivity artificial forage for ruminants (Table 4.45).

Table 4.45 Productivity of algae as artificial forage for ruminants

Protein source	*Protein yield (kg per hectare per annum)*	*Optimum growth temperature (°C)*	*Equivalent beef protein yield (kg per hectare per annum)*
Natural pasture	670		60
Spirulina platensis	24,350	27	2180
Chlorella pyrenoidosa (strain TX 7/11/05)	44,700	30	4030

Algal feed supplements (e.g. *Spongiococcus*) have sometimes been fed to poultry or cattle to improve the carotene pigmentation of eggs or milk.

4.5 Future prospects

For the immediate future, the new microbial proteins produced in quantity by growing yeasts on hydrocarbons or carbohydrates, or by growing bacteria on methane or methanol, are likely to find their main use as animal feed supplements. Some products are already slightly cheaper than soya-bean protein (the cheapest natural protein feed at present) and production costs may be expected to fall with improvements in technology. The immediate application of microbial protein production must be seen as a step in the cheaper production of meat, poultry, milk and eggs. As such, it will mainly benefit those countries which are sufficiently developed to have an intensive livestock raising system operating on a rational method of dietary allocation. Safety being assumed, in this context the only important factors are economic and nutritional: microbial feed proteins must be cheaper than, and of dietary value at least comparable with fish meal and soya meal.

The picture is complicated by the presently increasing price of a major substrate, natural mineral oil. Oil producing countries will no doubt set up their own plants for production of oil-derived feed proteins, to improve living standards by increased poultry and meat production. Oil-deprived countries may look to methanol, natural gases or other substrates to do the same. Others are investigating *Spirulina* and other algae (Mexico, in conjunction with the French Petroleum Institute, Czechoslovakia, USSR and Japan). Many units are no more than pilot scale, but some major projects are listed in Table 4.46.

It is difficult to see how underdeveloped countries, where severe protein-calorie malnutritions mainly occur, could benefit without outside help, for they lack not only the technology, capital and raw materials to set up and operate protein production plants, but also the high-productivity livestock industry which stands to gain most from a marginally cheaper protein source. Aid in many forms will be required.

Possible use of the new microbial proteins as a regular dietary supplement for humans is being approached with great caution by the major producers: long experience of successful use in farm animals, without evidence of toxicity, teratogenicity or carcinogenicity is felt to be necessary before formulations for human nutrition could seriously be contemplated. Moreover, there are two special problems. Acceptability is likely to prove a major difficulty, in developed and undeveloped countries alike, and there will be room for much ingenuity in first purifying the product so that it loses the characterstcics of the original microbial source, and then modifying it by flavouring and texturing so that it can simulate or even supplant existing foodstuffs. The other problem is the high nucleic acid content of microbial cells. None of the present commercial animal feed products is treated to reduce nucleic acid content, but this would be essential if regular human consumption were contemplated.

Table 4.46 Some microbial protein projects

Producer	Organism	Substrate
Dai Nippon Ink and Chemicals, Japan	Yeast	*n*-paraffins
Kanegafuchi Chemical Takasago Plant, Japan	Yeast	*n*-paraffins
Kyowa Kakko Kogyo, Japan	Yeast	*n*-paraffins
Mitsubishi Chemical Company, Japan	Yeast	petroleum fractions
Scottish BP (Grangemouth)	Yeast	*n*-paraffins
BP Lavéra, France	Yeast	gas-oil
BP/ANIC, Italy (Italproteine)	Yeast	*n*-paraffins
Regional Research Laboratories, India	Yeast	gas-oil
Chinese Petroleum Corporation, Formosa	Bacteria	fuel oil
ICI, Billingham, UK	Bacteria	methane
Rank Hovis McDougall, UK	Filamentous fungi	carbohydrates
Tate and Lyle, UK	Filamentous fungi	plant wastes

Fractionation would entail expensive cell rupture followed by equally expensive physical, chemical or enzymic removal of nucleic acids, and would greatly increase the cost of the final product. Genetic methods including DNA recombinant techniques, hardly used at all so far for food microorganisms, will doubtless be employed to produce strains combining such desirable properties as rapid growth, stability, non-toxicity, high intrinsic acceptability, correctly tailored amino acid and vitamin profiles, and even, perhaps, reduced total nucleic acid content.

Bibliography

AUTRET, (1972), 'Les protéines non conventionnelles, *C. R. Acad. Agr.*, **58**, 691–710.

BHATTACHARJEE, J. K., (1970), 'Microorganisms as potential sources of food', *Adv. Appl. Micribiol.*, **13**, 139–159.

COONEY, C. L. & LEVINE, D. W., (1972), 'Microbial utilization of methanol', *Adv. Appl. Microbiol.*, **15**, 337–363.

DAVIES, S. L. & WHITTENBURY, R., (1970), 'Fine structure of methane and other hydrocarbon-utilizing bacteria', *J. Gen Microbiol.*, **61**, 227–232.

DJAVANMARD, J. & GATELIER, C., (1969), 'L'application de la biochemie des hydrocarbures à la synthèse de nouvells sources alimentaires', *Rev. Techn. Petrole*, **95**, 39–47.

HARRISON, D. E. F., WILKINSON, T. G., WRENS, S. J.& HARWOOD, J. H., (1976), 'Mixed bacterial cultures as a basis for continuous production of SCP from C_1 compounds', in *Continuous Culture 6. Applications and New Fields*, Ed. A. C. R. Dean, D. C. Ellwood, C. G. T. Evans and J. Melling, Ellis Horwood.

HARRISON, J. S., (1971), 'Yeast production', *Prog. Ind. Micribiol.*, **10**, 129–179.

KIHLBÉRG, R., (1972), 'The microbe as a source of foods,' *Ann. Rev. Microbiol.*, **26**, 427–466.

KOSARIC, N. & ZAJIC, J. E., (1974), 'Microbial oxidation of methane and methanol', *Adv. Biochem. Eng.*, **3**, 89–125.

MATELES, R. I. & TANNENBAUM, (1968), *Single Cell Protein*, MIT Press.

MCLAREN, D. S., (1974), 'The great protein fiasco', *Lancet*, 93–96.

NOYES, R., (1969), *Protein Food Supplements*, Moyes Develop. Corporation, Park Ridge, U.S.A.

READE, A. E., SMITH, R. H. & PALMER, R. M., (1972), 'The production of protein for non-ruminant feeding by growing filamentous fungi on barley', *Biochem. J.*, **127**, 32.

REED, G. & PEPPLER, H. J., (1973), *Yeast Technology*, The AVI Publishing Company, Inc.

SCHELEGEL, H. G. & LAFFERTY, R. M., (1971), 'The production of biomass from hydrogen and carbon dioxide', *Adv. Bioch. Eng.*, **1**, 143–168.

SHACKLADY, C. A., (1974), 'SCP from hydrocarbons as animal feed ingredients', *Process Biochem.*, **9**, Dec., 9–11.

SHENNAN, J. L. & LEVI, J. D., (1974), 'Growth of yeasts on hydrocarbons', *Progr. Ind. Microbiol.*, **13**, 1–58.

VINCENT, W. A., (1971), 'Algae and lithotrophic bacteria as food sources', In *Microbes and Biological Productivity*, ed. D. E. Hughes and A. H. Rose, *Symp. Soc. Gen. Microbiol.*, **21**, 47–76.

VOLISKY, B. & ZAJIC, J. E., (1971), 'Batch production of protein from ethane and ethane-methane mixtures', *Appl. Microbiol.*, **21**, 614.

WILKINSON, J. F., (1971), 'Hydrocarbons as a source of single cell protein', In *Microbes and Biological Productivity*, ed. D. E. Hughes and A. H. Rose, *Symp. Soc. Gen. Microbiol.*, **21**, 15–46.

5 Products of Microbial Metabolism

The compounds produced by microorganisms may be grouped into three major categories: primary metabolites, secondary metabolites and enzymes.

5.1 Primary Metabolites

These are compounds involved in the biochemical pathways of actively growing microorganisms. They include the alcohols, ketones, organic acids, amino acids, nucleotides and polysaccharides. Some primary metabolites may be produced in concentrations far higher than those required by the producing organism (Table 5.1).

Table 5.1 Excessive production of certain primary metabolites

Products	Requirement for optimum growth (mg/l)	Production (mg/l)	Excess yield $\left(\dfrac{\text{Quantity produced}}{\text{Quantity required}}\right)$
Glutamic acid	300	60,000	$2{\cdot}0 \times 10^2$
Lysine	250	42,000	$1{\cdot}7 \times 10^2$
Inosinic acid (IMP)	25	13,000	$5{\cdot}2 \times 10^2$
Riboflavin	0·5	5000	$1{\cdot}0 \times 10^4$
Cyanocobalamine	0.001	30	$3{\cdot}0 \times 10^4$

Polyols, alcohols and ketones

Some strains of yeasts which normally produce ethyl alcohol by using a C_2 intermediate as a hydrogen acceptor can be induced, by making the C_2 intermediate unavailable, to use an alternative C_3 intermediate producing glycerol. Such a process is referred to as a steered fermentation. In fact, production of glycerol by a steered alcoholic fermentation is not profitable, because of the low yield, of the order of 20% based on glucose used, and extraction is difficult; also, a high concentration of sodium sulphite (40 g/l) is required for steering the fermentation. A number of osmophilic yeasts can, during growth in strongly sugary media, produce high yields of polyols. Yeasts of this kind can be isolated from the nectar of flowers and include *Saccharomyces rouxii, Saccharomyces mellis* and members of the genera *Hansenula, Pichia* and *Torulopsis*.

Another group of osmophilic yeasts can tolerate high concentrations of salt. The most common belong to the genus *Debaryomyces* and are found in the pickling brines used for such foods as meats, fish or gherkins, and also in certain food fermentations of the Far East, such as the production of miso, a commodity obtained from soya bean paste using *Pichia miso*.

Osmotic pressures tolerated by such yeasts depend, to some extent, on the substances involved (Table 5.2).

Table 5.2 *Saccharomyces rouxii*: growth-limiting concentrations of various substances, and corresponding osmotic pressures

Substance	Limiting concentration (%)	Osmotic pressure (atmos)
Glucose	80	220–260
Sucrose	80	180–200
Sodium chloride	20	150–165

In fact, many of these yeasts have a higher rate of growth on ordinary media than on media rich in sugars or salt, and are thus not fully osmophilic, but rather osmotolerant. Thus *Saccharomyces rouxii* grows well in sucrose concentrations between 5–50 % but only feebly at 80 %. It is of interest that the enzymes of salt-tolerant yeasts are not necessarily themselves salt-tolerant. For example, *Pichia miso* will ferment glucose to glycerol in the presence of 18 % sodium chloride: cell free extracts of the organism also yield glycerol from glucose but only at lower salt concentration.

Osmotolerant yeasts can be used for the production of polyols; in general these metabolites are best formed in the presence of a readily utilized nitrogen source, such as urea or ammonium salts, with carefully controlled aeration, relatively high temperature (37°C) and low phosphate concentration; glucose concentration may be in the region of 30 %. *Torulopsis magnoliae* produces glycerol in yields of up to 50 % based on glucose consumed: *Hansenula subpelliculosa* produces D-arabitol. A strain of *Pichia miso* produces a mixture of D-arabitol, erythritol and glycerol; in the presence of more than 6 % sodium chloride it produces glycerol only.

Although these particular polyols are not widely used, osmophilic yeasts have been investigated as agents for producing new, potentially useful products. Thus xylitol, a possible replacement for glucose in the diets of diabetics, could be obtained by direct chemical reduction of D-xylose but this would be very expensive. No known microorganism produces xylitol directly from glucose but a three stage fermentation has been devised (Table 5.3) in which D-arabitol and D-xylulose are intermediates. There is no need to isolate the intermediates at each stage. In this process, a medium containing 15% glucose yields a solution containing 1·8% xylitol (an 11·6% yield in terms of glucose consumed). Heat sterilization after the first stage destroys the *Debaryomyces* cells, giving a source of amino acids and growth factors for *Acetobacter suboxydans* (= *Gluconobacter oxydans* subsp. *suboxidans*) in the second stage.

A multistage process could also be employed for the biosynthesis of D-fructose, used in the food industry because of its sweetness and high solubility. The starting material could be glycerol, which is becoming increasingly available from the petrochemical industry; it could be converted first to mannitol and subsequently to D-fructose (Figure 5.1).

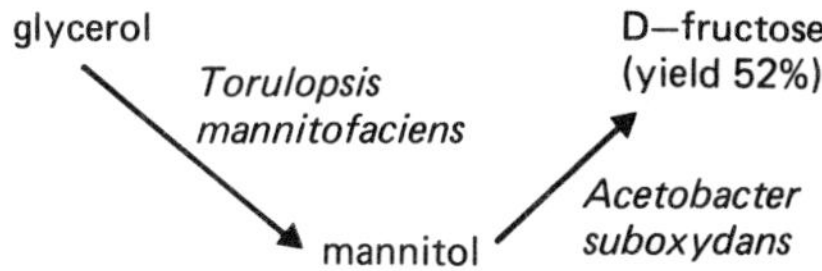

Figure 5.1   Production of D-fructose from glycerol

Dihydroxyacetone can also be obtained from the oxidation of glycerol by a strain of *Acetobacter suboxydans*, in a medium containing 0·5% corn steep liquor at pH 5·5. The process is capable of yields as high as 90% of substrate. Dihydroxyacetone is used in certain cosmetics to produce an artificial browning of the skin, which it does by reacting with arginine groups in the skin proteins. It has also been shown to be useful in the treatment of diabetes, by preventing the formation of acetoacetic acid and β-hydroxybutyrate.

3-ketomaltose, and antioxidant used in the food industry for the protection of lipids, can be obtained from maltose by the action of *Agrobacterium tumefaciens*.

Industrial solvents

Ethanol, glycerol, butanol, acetone and 2,3-butanediol have all been produced by microbial fermentation but microbiological processes have tended to be supplanted by chemical ones for two reasons,

(i)   increase in price of the fermentable organic substrates
(ii) increasing economic competition from the petrochemical industry which

Table 5.3 Production of xylitol by fermentation (the medium used also contains: KH_2PO_4 (0·1 %); $MgSO_4.7H_2O$ (0·01 %); $CaCl_2.2H_2O$ (0·01 %); NaCl (0·01 %) in distilled water)

| | Medium | | | Incubation | | | | Yield % | |
Step	Composition (%)	pH	Conditions of sterilization	Time (h)	Temp. (°C)	Organism	Product	on previous substrate	overall on glucose
1	Cornsteep 4 Glucose 15	5	5 min 110°C	99	30	Debaryomyces hansenii	D-arabitol	34·2	34·2
2	—	6	15 min 120°C	48	30	Acetobacter suboxydans	D-xylulose	94·3	32·8
3	Cornsteep 4	6	5 min 110°C	54	30	Candida guilliermondii	xylitol	41·8	11·6

produces such solvents as byproducts of the refining of petroleum. However, some microbiological processes for the production of industrial solvents are outlined in Table 5.4.

Organic acids

Various organic acids may accumulate in microbial cultures as the result of metabolism of carbohydrates. They may be terminal products of glycolysis (lactic and propionic acids) or they may result from the incomplete oxidation of sugars (citric, itaconic, gluconic acids). Another type of product is acetic acid, obtained from the dehydrogenation of alcohol in the presence of oxygen.

Citric acid

The production of citric and itaconic acids are closely related. These compounds accumulate in the medium as a result of the interruption of the Krebs' cycle (Figure 5.2); citric acid is a key intermediate in the cycle: itaconic acid is formed by decarboxylation of *cis*-aconitic acid.

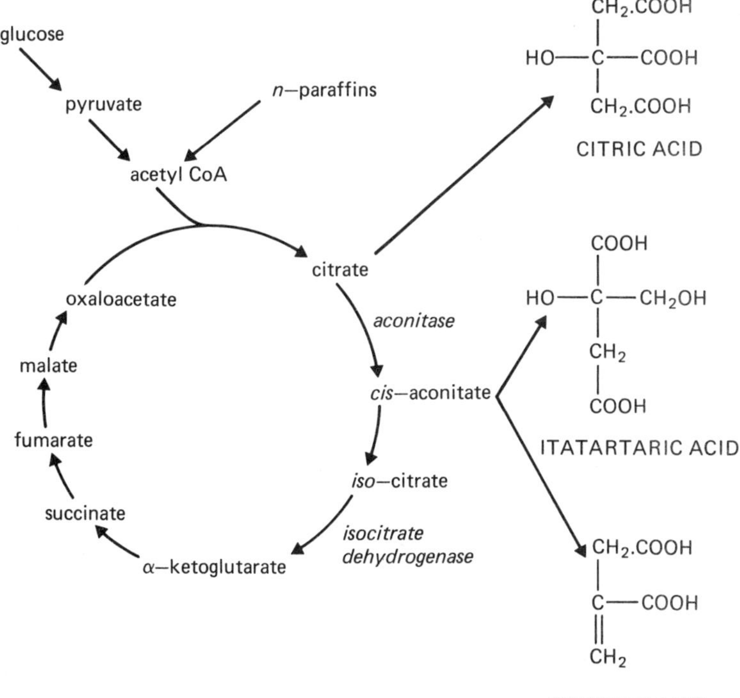

Figure 5.2   Tricarboxylic acid cycle (Kreb's cycle)

Table 5.4 Production of industrial solvents by fermentation

Product	Organism	Substrate	Medium Constituents	pH	Time and temp	Yields	Use
Ethanol	*Saccharomyces cerevisiae* *Candida pseudotropicalis* *Saccharomyces cerevisiae* *Candida utilis*	Molasses Milk whey Starch hydrolysates Sulphite liquors	Phosphates $(NH_4)_2SO_4$	4·0–4·5	48–72 h 30°C	65 l alcohol per 100 kg sucrose	Solvent/ precursor of many synthetic chemicals
Butanol + Acetone	*Clostridium acetobutylicum*	Molasses, starch, inulin, sulphite liquors, cellulose hydrolysate	Phosphate $(NH_4)_2SO_4$ NH_3	5·0–7·0	48 h 30–37°C	28 kg butanol 10 kg acetone 1 kg ethanol per 100 kg sugar	Solvent/ explosive (cordite)
2,3-Butane-diol	*Bacillus polymyxa* *Bacillus subtilis* *Enterobacter aerogenes* *Aeromonas hydrophila*	Molasses, starch, cellulose hydrolysates	Phosphate $(NH_4)_2SO_4$				Synthesis of rubber

More than 100,000 tonnes of citric acid are manufactured annually throughout the world. It is widely used in the food industry (fruit drinks, confectionery, jams, preserved fruits), cosmetics (astringent lotions) and others. The ability of citric acid to sequester metal ions in acidic media makes it useful in the formulation of shampoos and electrolyte solutions, in the tanning of animal skins, and even for the clearing of pipes in the oil industry.

A strain of *Aspergillus niger*, selected after a programme of mutations and hybridizations, and capable of producing up to 80 g or more of citric acid per 100 g glucose, is used. Strains may initially be screened for acid production by inoculating them onto sheets of filter paper impregnated with a sucrose medium containing a pH indicator. After incubation, the diameter of the zone of colour change gives an indication of the degree of acid production, allowing selection of high-yielding strains. The process occurs in two stages, the first being the rapid development of mycelium, whereas during the second phase growth ceases and the required acid accumulates.

Citric acid synthesis occurs when the medium is unbalanced. It is presumed that a constituent essential for the functioning of one of the enzymes of the tricarboxylic acid cycle is missing. Certainly, aconitase requires iron as a cofactor and copper acts as an antagonist of iron, blocking the action of aconitase. This enzyme is found in actively growing cells but is no longer active during the phase of citric acid production.

The industrial production of citric acid is surrounded by a certain amount of secrecy but some of the optimum conditions, at least as indicated by laboratory studies, appear to be as follows.

The *Aspergillus niger* strain must be grown as a surface culture or in a well aerated fermenter. The inoculum is obtained by growing the mould on a solid medium at 25°C for 4–14 days and suspending the resulting spores in water.

Table 5.5 Effect of iron concentration on citric acid production by a mutant of *Aspergillus niger*

Iron (mg/l)	*Yield of citric acid*[*]
0·0	67
0·05	73
0·50	88
1·00	76
10·0	39

$$^{*}\left(\frac{\text{wt of acid produced}}{\text{wt of glucose used}} \times 100 \right)$$

The pH of the production medium must be less than 3·5, obtainable by including an ammonium salt of a strong acid, to avoid formation of oxalic and gluconic acids. Sucrose concentration should be of the order of 14%, and the availability of nitrogen restricted to prevent the formation of oxalic acid (0·25% NH_4NO_3). Phosphate concentration should be less than 0·25%. It is essential to control the concentration of iron (Table 5.5) and copper (Table 5.6).

If the sugar substrate is molasses, ferrocyanide can be added to the medium before sterilization, to precipitate excess iron, or the clarified molasses can be passed through an ion exchange resin. Under laboratory conditions, highest yields are obtained using sucrose passed through an ion-exchange resin (Table 5.7). A defined medium based on sucrose, purified in this way, can be used to determine optimum concentrations of the various trace elements. Using a highly purified carbon source, it is necessary to add copper and zinc (0·1–50 ppm) to the medium.

Because of the low pH, and the sensitivity of the process to excess iron, the fermentation vessels must be protected from corrosion or constructed of stainless steel. High aeration rates are essential (0·5–1·5 volumes per minute), producing much foam; fermentation takes 5–14 days at 27–33°C.

Table 5.6 Effect of copper concentration on yields of citric acid from *Aspergillus niger*

Iron (ppm)	Copper (ppm)	Yield of citric acid*
10	50	77·8
50	50	69·1
100	50	50·7
150	50	14·2
10	100	77·2
50	100	65·4
100	100	53·9
150	100	29·8

$$* \left(\frac{\text{wt of acid produced}}{\text{wt of glucose used}} \times 100 \right)$$

Isolation of the citric acid starts with its precipitation from the fermented broth using lime. The precipitate is washed, treated with sulphuric acid to remove calcium as the very insoluble sulphate, and the resulting dilute solution of citric acid passed through a column of carbon granules. The purified solution is then concentrated under vacuum until crystallization occurs.

Table 5.7 Effect of substrate purity on production
of citric acid by *Aspergillus niger*

Treatment	Yield of citric acid*
Sucrose passed through resin	98
Molasses clarified and passed through resin	75
Molasses treated with ferrocyanide	68
Molasses untreated	62

$$* \left(\frac{\text{wt of acid produced}}{\text{wt of glucose used}} \times 100 \right)$$

Alternatively, counter current extraction with organic solvents (100 parts of tri-*n*-butyl phosphate and 5–30 parts *n*-butyl acetate) may be used. Citric acid is then extracted from the organic phase with lime water and the precipitate of calcium citrate treated with sulphuric acid as previously described.

Certain strains of *Candida lipolytica* produce a mixture of citric and isocitric acids from normal paraffins. Monofluoroacetate is readily incorporated into the tricarboxylic acid cycle, in place of acetate, giving rise to monofluorocitrate which is a powerful competitive inhibitor of *cis*-aconitase. By adding monofluoroacetate to media Japanese workers have been able to isolate mutants with only one hundredth of the *cis*-aconitase activity shown by the parent culture (*Candida lipolytica* ATCC 20114). Such strains are capable of producing 112 g of citric acid per litre after 3 days incubation at 30°C representing a yield of 145 % based on substrate consumed.

Itaconic acid

This compound is used in the manufacture of plastics: together with its esters, it is readily polymerized, being essentially a substituted acrylic acid. A number of interesting plastics are obtained by copolymerizing itaconic acid with other monomers such as acrylic acid, methylacrylate or styrene. Itaconic acid may also be used as a material for artificial dentures and for certain synthetic fibres such as Courtelle. Mutants of an *Aspergillus terreus* strain (NRRL 1960), isolated from a Texan soil, are still used in the industrial production of itaconic acid.

Production of itaconic acid, like citric acid, is sensitive to medium composition; concentrations of iron (Table 5.8), copper and zinc (Table 5.9) and calcium (Table 5.10) are all important.

Table 5.8 Effect of iron
concentration on itaconic
acid production by a mutant
of *Aspergillus terreus*

Iron (mg/l)	*Yield**
0	57
1	25
2	17
4	17

$$*\left(\frac{\text{g of itaconic acid}}{\text{g of hexose used}} \times 100\right)$$

Table 5.9 Effect of copper and zinc
concentration on itaconic acid
production by a mutant of
Aspergillus terreus

Copper (mg/l)	*Zinc* (mg/l)	*Yield**
0	0	16
0	0·5	43·3
0	6·0	50·4
0·5	0·5	55·4
1	0·5	51·7
3	0·5	52·6
6	0·5	55·4

$$*\left(\frac{\text{g of itaconic acid}}{\text{g of hexose used}} \times 100\right)$$

A competing reaction in the production of itaconic acid is the synthesis of itatartaric acid which is an undesirable byproduct of many itaconic acid producing strains of *Aspergillus terreus*. Calcium inhibits the enzyme involved in the production of itatartaric acid thus releasing more substrate for itaconic acid production (Table 5.10).

Free itaconic acid is toxic to moulds at concentrations in excess of 7% but, by step-wise addition of ammonia to neutralize the acid, concentrations of up to 20% may be accumulated in the medium.

A typical medium contains about 15% of sugar, either pure or in the form of molasses. Fermentation takes 3 days at 32–35°C; after filtering off the mycelium, the solution is concentrated until itaconic acid crystallizes; the acid can also be extracted using amyl alcohol.

Table 5.10 Effect of calcium
concentration on itaconic acid
production by a mutant of
Aspergillus terreus

Calcium (mg/l)	Yield*
0	9
337	43
2700	59

$$* \left(\frac{\text{g of itaconic acid}}{\text{g of hexose used}} \times 100 \right)$$

A number of other organic acids can be produced by fermentation: some are listed in Table 5.11 which gives comparative process details. Citric acid and itaconic acid have been discussed in some detail as exemplifying typical problems in the production of organic acids by fermentation.

Vitamins

Prototrophic microorganisms can synthesize all of the vitamins, coenzymes, and other complex growth factors needed for their growth and metabolism from simple compounds in the culture medium. These factors are generally elaborated in amounts just sufficient for the organism's needs, and do not accumulate in the culture medium. A few organisms, when grown under highly specified and artificial conditions, produce vitamins or other growth factors greatly in excess of their own metabolic needs, and are commercially interesting. A few vitamins are manufactured commercially (Table 5.12).

Vitamin B_{12}

This complex compound is essential for normal growth in man and domesticated animals; in man the daily requirement is 0·001 mg per day. In pernicious anaemia in man, absorption of vitamin B_{12} is impaired, due to changes in certain secretory cells of the stomach, and the disease is treated by large dietary supplements or, better, by intramuscular injections. The vitamin is produced by the microflora of the rumen; it accumulates in the liver of ruminants, but in such a low concentration (c 1 ppm) that extraction is not feasible. Vitamin B_{12} is also produced by the intestinal microflora of man and other non-ruminant animals, but intestinal absorption is ineffective and most of the vitamin is excreted in the faeces. Some herbivores, such as the rabbit, are coprophagous: these absorb the vitamin present in their faeces *via* the stomach. Other mammals, including man, depend mainly on animal products

Table 5.11 Other organic acids produced by fermentation

Product	Culture	Substrate	Process	Extraction	Yield (%)	Uses
Acetic acid	*Acetobacter* sp.	Ethanol	Continuous aerated process using an alcohol solution containing (%): glucose 0·9 ammonium phosphate 0·4 magnesium sulphate 0·1 potassium citrate 0·1 pantothenic acid 0·00005	Filtration	98–99	Food industry
Lactic acid	*Lactobacillus delbrueckii*	Milk whey Molasses Pure sugars	10–15% glucose, 5–6 days, 50°C in corrosion resistant fermenter pH5·5–6·0 buffered with $CaCO_3$ No aeration, growth factors provided by malt	Precipitation after heating to 80°C and the addition of chalk (calcium lactate is formed) —extraction with solvents —esterification with methanol followed by distillation	90	Food industry (fruit extracts, lemonade, syrups, pickles) Dye mordant Tanning, decalcifying skins Plastics Medical (iron and calcium lactates)

Propionic acid	*Propionibacterium*	Lactose Glucose Sucrose Starch	8–12 days at 30°C without aeration Yeast extract as a source of growth factors			Industrial (thermoplastics, perfumes, fungicide formulation)
Fumaric acid	*Rhizopus* sp. *Candida hydrocarbo- fumarica*	Glucose *n*-alkanes	3 days, 30°C Aeration, pH5–6 maintained by the addition of NaOH	Acidification of media and crystallization	60	Resins
Gluconic acid	*Aspergillus niger*	Glucose and corn steep liquor	pH6·5, aeration 36 h at 30°C	Sodium gluconate is obtained after filtration and passing through a cationic ion exchange resin	90	Medical (calcium, iron and potassium gluconates) cleaning (metals, bottles) Food acid

Table 5.12 Production of some vitamins using microorganisms

Vitamin	Culture	Medium	Fermentation conditions	Extraction	Yield
Carotene (precursor of vitamin A)	*Blakeslea trispora*	Molasses, soya oil, β-ionone, thiamine	72 h at 30°C aerobic	Solvent	1 g/l
	Mycobacterium smegmatis				0·007 g/l
Riboflavin	*Ashbya gossypii*	Glucose, collagen, soya oil, glycine	6 days 36°C aerobic	Heated 1 h at 120° followed by addition of reagents to precipitate riboflavin	4·25 g/l
L-sorbose (in Vitamin C synthesis)	*Gluconobacter oxidans* subsp. *suboxidans*	D-sorbitol, 30% corn steep	45 h at 30°C aerobic	Filtration and concentration under vacuum	70% based on substrate used
5-keto gluconic acid (in vitamin C synthesis)	*Gluconobacter oxidans* subsp. *suboxidans*	Glucose, $CaCO_3$, corn steep	33 h at 30°C aerobic	Filtration and concentration under vacuum	100% based on substrate used

Vitamin B_{12} see Table 5.13.

as a source of the vitamin. Pigs and poultry grow well on a vegetable protein diet only if 10–15 mg of vitamin B_{12} is added to each ton of feed.

In nature, therefore, biosynthesis of vitamin B_{12} is entirely microbial. Various bacteria and streptomycetes are used to produce it commercially for pharmaceutical or feed-supplement use, and yields of up to 20 mg/l are obtainable. Early methods of purification involved difficult and time-consuming chromatography; at present, the extraction method is similar to that used for streptomycin and aureomycin. Highest yields have been obtained using strains of *Propionibacterium* (Table 5.13).

Vitamin B_{12} is cyanocobalamin, consisting of a molecule of cobinamide linked to a nucleotide. The cobinamide molecule has a central atom of cobalt linked to a cyanide group, surrounded by four reduced pyrrole rings joined to form a macro-ring; a number of the carbon atoms carry methyl or other substituent groups. The nucleotide is atypical in having 5,6-dimethyl benz-iminazole as its base instead of a purine or pyrimidine base (Figure 5.3).

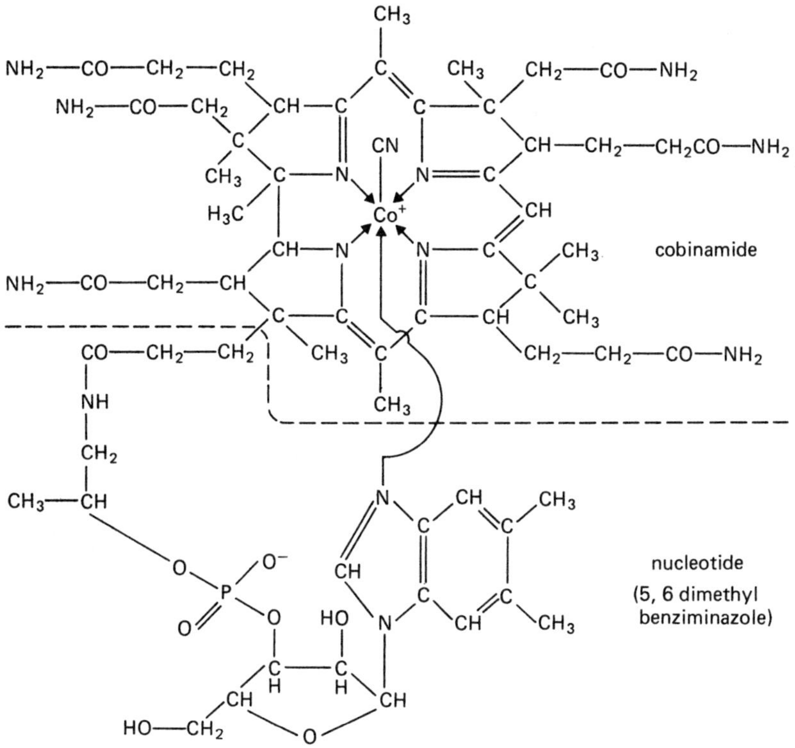

Figure 5.3   Vitamin B_{12}: Cyanocobalamin

In production of vitamin B_{12} using *Propionibacterium freudenreichii*, 0·5 % betaine is included in the medium to furnish methyl groups which it probably does by aiding in the synthesis of methionine which acts as a methyl donor. The concentration of cobalt in the medium is critical, 5 μg/ml being optimum; too much suppresses biosynthesis. Cobinamide formation is inhibited by oxygen: there is thus a preliminary anaerobic fermentation of 70 hours, during which cobinamide is produced and the pH falls from 7·5 to 6·5 and then rises to 8·5. 0·1 % 5,6-dimethylbenziminazole is added, since *P. freudenreichii* cannot readily synthesize this base, and during a further 50 hours aerobic fermentation the nucleotide is formed and links with the cobinamide to give cobalamin (yield <20 μg/ml). The cobalamin is endocellular and is released from the centrifuged cells by acidifying to pH 2–3 and briefly heating to 100°C. After removal of cell debris, potassium cyanide (5 μg/ml) is added to convert cobalamin to cyanocobalamin and also sodium sulphite to prevent oxidation of the final vitamin. In a continuous culture version of the process, two large fermentation vessels in series are used, each having a retention period of 60

Table 5.13 Production of vitamin B_{12} by microorganisms

Species	Medium	Aeration	Temp	Time of process (h)	Yield (mg/l)
Bacillus megaterium	Molasses, mineral salts, cobalt	aerobic	30°C	18	0·45
Propionibacterium freudenreichii	Glucose, corn steep, betaine, cobalt pH 7·5	anaerobic (3 days) + aerobic (2 days)	30°C	120	20
Propionibacterium shermanii	Glucose, corn steep, ammonia, cobalt pH 7·0	anaerobic (3 days) aerobic (4 days)	30°C	150	23
Streptomyces olivaceus	Glucose, soya flour, cobalt, mineral salts	aerobic	28°C	96	5·7
Bacillus coagulans	Citric acid, triethanolamine, cobalt, corn steep	aerobic	55°C	18	6·0
Pseudomonas denitrificans	Oxalic acid, betaine, cobalt, mineral salts	aerobic	—	—	10

hours: the first is anaerobic and the second aerobic, with 5,6-dimethyl-benziminazole added continuously: the yield is $15\mu g/ml$.

The yield of vitamin B_{12} is assayed microbiologically using appropriate B_{12} deficient strains such as *Lactobacillus leichmannii* or *L. lactis*.

Japanese workers have shown that various strains of *Corynebacterium*, *Mycobacterium* and *Nocardia* can produce vitamin B_{12} in hydrocarbon media: yields are only of the order of a few milligrams per litre.

Amino acids

Many microorganisms can synthesize their amino acid requirements for protein biosynthesis from glucose and ammonium salts. Amino acids are only transitory intermediates in metabolic processes, but small quantities may be released into the medium at the end of the exponential growth phase. Such release of amino acids may be much higher if the organism has lost one or more of the normal mechanisms of regulation.

Patents have been produced (especially in Japan) describing production of most of the naturally occurring amino acids, but only glutamic acid and lysine are manufactured in large quantities; both are used principally in the food industry.

Glutamic acid

The world production of glutamic acid is more than 100,000 tonnes per year. As mono-sodium glutamate it is used in the food industry as a flavour ingredient in such foods as soups.

Glutamic acid can be produced in two stages, one culture producing α-keto glutaric acid, another converting this to glutamic acid. However, a single-stage process is mainly used: Japanese workers have obtained, from soil isolates of *Corynebacterium glutamicum*, mutants producing yields of up to 60 g/l (the organism itself requires no more than 300 mg/l). The fermentation takes 40 hours at 30°C; the medium is slightly alkaline and should contain only traces

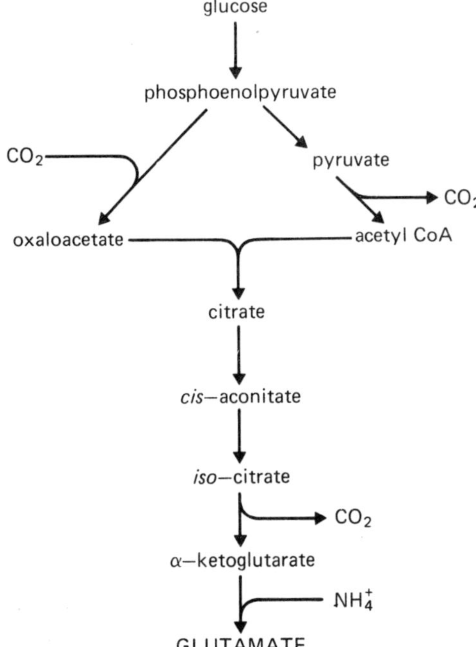

Figure 5.4 Principal intermediates in the production of glutamic acid from glucose

of biotin $(1-5\,\mu g/l)$. If penicillin or detergents are included in the medium, glutamate formation occurs even at higher biotin concentrations, and pure glucose can be replaced with molasses. Yields on molasses are higher than might be expected because molasses contains organic acids (e.g. acetic and citric) which may also be converted into glutamic acid.

Corynebacterium glutamicum metabolizes sugars *via* the glycolytic pathway and not *via* the monophosphate shunt so production of glutamate ceases in the presence of classical inhibitors of glycolysis such as fluoride and iodoacetate. Accumulation of glutamic acid takes place because the mutant lacks an α-ketoglutarate dehydrogenase and thus has an incomplete tricarboxylic acid cycle (Figure 5.4). Provision of energy is maintained by a functioning glyoxylate pathway (Figure 5.5).

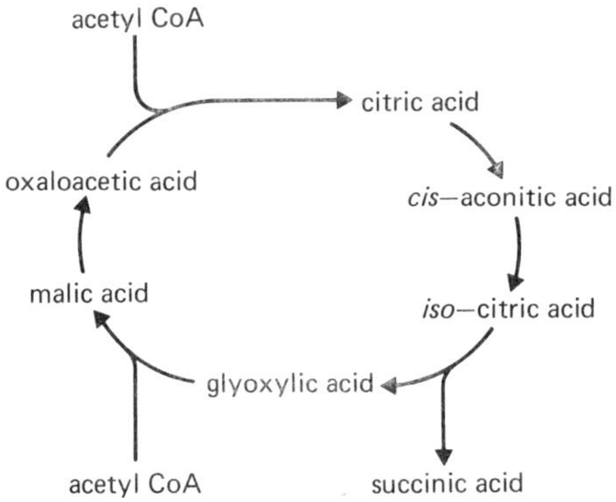

Figure 5.5   The glyoxylic acid cycle

The α-ketoglutarate derived from the tricarboxylic acid cycle is reductively aminated by a glutamate dehydrogenase utilizing $NADPH_2$, itself formed by the dehydrogenation of *iso*-citrate (Figure 5.6).

During the growth phase glutamate accumulates in the cells until a saturation point is reached (50 mg/g dry weight of cells). Subsequently, as a result of biotin deficiency, the cells become permeable to glutamate and it is excreted into the medium. If a medium containing molasses (which has ample biotin) is used, addition of penicillin or detergents during the exponential growth phase allows glutamate release.

Corynebacterium hydrocarboclastus strain R.7 metabolizes hexadecane and, in the presence of either penicillin or Tween 80, releases glutamate with yields of up to 75 g/l; the medium must contain small amounts of thiamine $(3-5\,\mu g/l)$, the coenzyme of α-ketoglutarate dehydrogenase, in the absence of which α-ketoglutarate accumulates rather than glutamate.

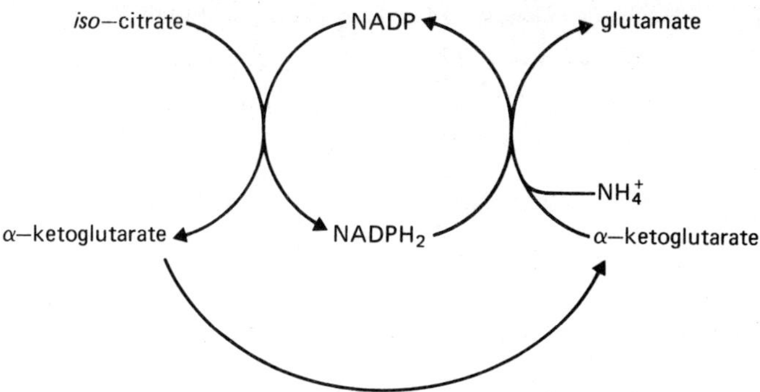

Figure 5.6   Reductive amination of α-ketoglutarate

In attempts to produce glutamic acid from hydrocarbons without the need to add penicillin, which complicates subsequent purification, Japanese microbiologists have investigated the use of mutants auxotrophic for glycerol. A mutant of *Corynebacterium alkanolyticum* 314, termed GL 21, lacked L-glycerol 3-phosphate—NADP oxidoreductase and excreted 70 g/l of glutamic acid after 80 hours at 30°C. This incredible yield was apparently due at least in part to modified cell permeability resulting from impaired ability to synthesize phospholipids.

A number of other amino acids can be produced by microbiological processes but only lysine is manufactured on a significant scale (see Figure 1.11).

Nucleotides

The industrial production of these substances has been the subject of extensive research, particularly in Japan. The three ribonucleoside 5′-monophosphates, guanylic acid (guanosine 5′-monophosphate or GMP), inosinic acid (inosine 5′-monophosphate or IMP) and xanthylic acid (xanthosine 5′-monophosphate or XMP) all have taste enhancing properties which reinforce the effect of mono sodium glutamate. They also give the illusion of a certain viscosity in soups creating an impression of body and rich flavour. Because of these properties studies have been made of their production by fermentation methods, both direct and indirect, usually involving suppression of regulatory mechanisms.

Direct fermentation

The biosynthetic pathway for the production of 5′-nucleotides is outlined in Figure 5.7. Mutants of *Corynebacterium glutamicum* or of *Brevibacterium*

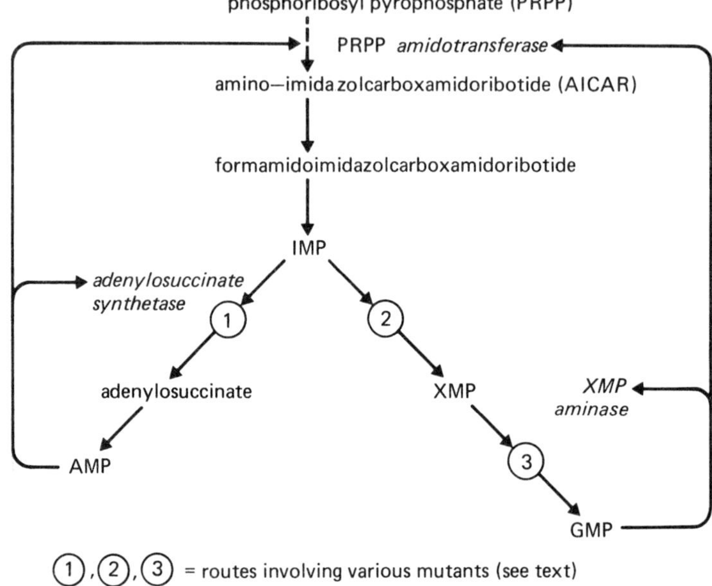

Figure 5.7   Biosynthesis of 5' nucleotides

ammoniagenes which require adenine (mutation (1) lacking adenylsuccinate synthetase) and xanthine (mutation (2) lacking IMP dehydrogenase) are used.

AMP and GMP control the pathway by feed-back inhibition and repression of the first enzyme, PRPP amidotransferase. If the medium contains limiting concentrations of adenine, xanthine and biotin, IMP may accumulate in it in concentrations of up to 20 g/l.

Auxotrophs for adenine and guanine which lack XMP aminase (mutant 3) may accumulate up to 70 g/l of XMP. To obtain GMP by direct fermentation is difficult, because GMP is a terminal product which controls PRPP amidotransferase by feed back inhibition and also inhibits IMP dehydrogenase. However, a selected revertant of an IMP dehydrogenase-lacking mutant was found to produce an IMP dehydrogenase no longer susceptible to feedback inhibition, and yielded up to 1 g of GMP per litre of growth medium.

The nucleotides synthesized by these high yielding strains are, of course, intracellular. The mechanism by which they are enabled to pass through the cytoplasmic membrane into the culture medium is unknown.

Some organisms, e.g. *Bacillus subtilis*, produce hydrolytic enzymes such as nucleotidases, nucleosidases and phosphatases, which are capable of degrading nucleotides and must therefore be controlled if nucleotide production is the aim. Addition of copper to the medium blocks extracellular nucleotidases and phosphate in the medium represses the formation of cell bound phosphatases; also, it is possible to obtain mutants deficient in cell-bound nucleosidase (e.g., one mutant of *B. subtilis* exhibits only 0.5% of the nucleosidase activity of the wild-type strain).

Strains of *Arthrobacter paraffineus* can produce an excess of nucleotides from hydrocarbons, and an adenine auxotrophic mutant produces up to 2·8 g of IMP per litre of medium. One adenine-requiring mutant of *Brevibacterium ammoniagenes* apparently synthesizes IMP extracellularly. In studies on this strain, it was found that hypoxanthine accumulated in the medium at the start of the fermentation. This base later disappeared and was replaced by IMP. It would seem that the strain possesses an intact pathway for intracellular biosynthesis of IMP: however, instead of being excreted intact, IMP is broken down to the corresponding base which is excreted. Resynthesis of IMP occurs outside the cell by the following reaction:

hypoxanthine + phosphoribosyl pyrophosphate (PRPP) →

$$IMP + pyrophosphate$$

This has suggested an alternative process for the manufacture of nucleotides using wild type strains. The corresponding base is added to a medium rich in phosphate, magnesium, thiamine and calcium pantothenate. The concentration of manganese has been shown to be critical. If it is low enough (less than 10 μg/l) the permeability of the cell membrane is altered in such a way as to allow excretion of PRPP and with carefully controlled conditions it is possible to bring about the extracellular biosynthesis of the required nucleotide.

Indirect fermentation

Microorganisms of many genera, *Pseudomonas*, *Achromobacter*, *Alcaligenes* and *Candida* can be grown using normal paraffins as their carbon source, the aim usually being to produce microbial foodstuffs. However, the product usually has an unduly and undesirably high nucleic acid content (see Chapter 4). RNA can be extracted from the harvested biomass, treated with a complex enzyme system, also of microbial origin (e.g. *Penicillium citrinum* or *Streptomyces* spp.), containing 5′-phosphodiesterases, and a mixture of 5′-nucleotides obtained.

Polysaccharides

Many microorganisms can synthesize extracellular polysaccharides: in liquid culture, the medium becomes more viscous; on agar-based media containing sugars, extracellular polysaccharide-producing colonies can be recognized by their slimy or viscous nature and are easily selected. The ability to produce extracellular polysaccharide may benefit an organism in a number of ways. Many polysaccharides are able to retain water, perhaps protecting the producing microorganism from desiccation. Polysaccharide capsules may protect bacteria from phages, from phagocytosis in an animal host, or from predation by protozoa in water or soil. Furthermore microbial extracellular polysaccharides (which are frequently quite resistant to biodegradation) may represent a useful carbon and energy reserve deposited outside the cell to avoid osmotic problems, and ready for extracellular enzymic hydrolysis when required.

It is possible to distinguish two major groups of polysaccharide: homopolysaccharides, such as dextrans, usually consisting of straight or branched chains of one type of sugar molecule only, and produced by a single enzyme, and heteropolysaccharides, containing a number of different kinds of sugar and other molecules, biosynthesis often requiring a number of different enzymes.

A specific carbon source is not necessarily required, but for high polysaccharide yield media should in general be relatively deficient in nitrogen and phosphorus. Aerobic conditions are always required: fermentation under anaerobic conditions does not generate sufficient biochemical energy for the synthesis of high yields of polysaccharide.

Microbial polysaccharides can frequently be produced from relatively cheap raw materials and are thus potentially useful substitutes for the water soluble vegetable gums. Industrial applications are numerous and varied: stabilizers in the food industry for such products as syrups and ice creams; pigment suspension in paints of high viscosity; dispersion of pigments for fabric printing and dying; additives to drilling mud in the oil exploration industry; extraction of crude oil from oil bearing shales; improvement of the viscosity of cosmetic lotions and dentifrices.

High yields, approaching 50% based on the carbon source, are obtainable from a number of bacterial genera including *Leuconostoc*, *Xanthomonas*, *Rhizobium*, *Arthrobacter*, *Bacillus* and *Acetobacter;* and from yeasts such as *Hansenula* and *Cryptococcus*.

Xanthans

Cultures of *Xanthomonas phaseoli*, responsible for a disease of haricot beans, produce an interesting polysaccharide, aqueous solutions of which have a viscosity which is independent of temperature, pH and the presence of mineral salts. Because of this unusual property, the substance is of value in formulations for drilling muds, especially in undersea oil exploration. *Xanthomonas campestris* NRRL B.1459 also gives a very high yield of a similar polysaccharide; after 96 hours at 30°C, 50% of the glucose in the medium is transformed into a complex heteropolysaccharide containing D-glucose, D-mannose and D-glucuronate in the molecular ratio of $2 \cdot 8 : 3 : 2$ and with a molecular weight of about 10^6 Daltons (Figure 5.8). There are also 3·4 moles of P-acetyl, 1·0 mole of pyruvate and 5·0 moles of K^+ and Na^+ as the salts of glucuronic and pyruvic acids.

$$[-GA(\beta.1{:}2)M(1{:}4)GA(\beta.1{:}2)M(1{:}4)G(\beta.1{:}4)G(1{:}4)]_n$$

$$M(1{:}2)$$

GA = glucuronic acid
M = D—mannose
G = D—glucose

Figure 5.8   Structure proposed for the basic unit of xanthan

For use as a marine drill lubricant, the polysaccharide is cheaply and crudely manufactured simply by drying whole liquid cultures of the bacterium. If required in a purer state, it may be precipitated with methanol or other solvents.

Other microbial polysaccharides are manufactured in varying quantities for a range of uses (Table 5.14). Dextrans are no longer used as widely as in previous times as a substitute for blood plasma.

Table 5.14 Microbial polysaccharides

Culture	Substrate	Product	Use
Zoogloea ramigera	Varied	Heteropolysaccharide	Absorption of impurities: (amino acids, toxic cations, radio-active cations)
Acetobacter aceti subsp. *xylinum*	Hexoses	Cellulose	Osmotic membranes
Leuconostoc mesenteroides	Sucrose	Dextrans	Substitute for blood plasma, iron complex (medical), molecular sieve (Sephadex)
Neisseria sp.	Sucrose	Starch	none
Hansenula sp.	Glucose	Phosphomannans	none
Arthrobacter sp.	Glucose	Acetylated heteropolysaccharide	none

5.2   Secondary metabolites

Under this heading, biological pesticides, gibberellins and alkaloids will be considered. Consideration of another most important group of secondary metabolites, the antibiotics, is reserved for a separate chapter (Chapter 6).

Biological pesticides

It has been shown that aqueous suspensions of some antibiotics (50–200 ppm) have insecticidal activity. Thus, novobiocin and cycloheximide (= actidione) have a wide spectrum of activity against different insect species, though it is not clear whether they act by ingestion or contact. Screening programmes have been carried out in Japan leading to the discovery of new metabolites from streptomycetes with more specifically insecticidal activity (Table 5.15).

Table 5.15 Insecticidal metabolites of microorganisms

Culture	Product	Toxicity to mammals
Streptomyces pactum	Pactomycin	High
Streptoverticillium mobaraense	Piericidins A and B	High
Metarrhizium anisopliae	Destruxins A and B	High
Aspergillus ochraceus	Aspochracin	Low
Aspergillus versicolor	Versimide	—

From higher fungi, traditionally used in Japan for controlling flies, two new amino acids (Figure 5.9) have been isolated, tricholomic acid from *Tricholoma muscarium* and ibotenic acid from *Amanita muscaria*. Tricholomic and ibotenic acids have, curiously, flavour-enhancing properties similar to those shown by the 5′-nucleotides, and ibotenic acid also possesses psychomimetic properties. This range of biological activity in relatively simple compounds has generated considerable interest. It is interesting to note the similarity between the structures of these compounds and that of the antibiotic cycloserine.

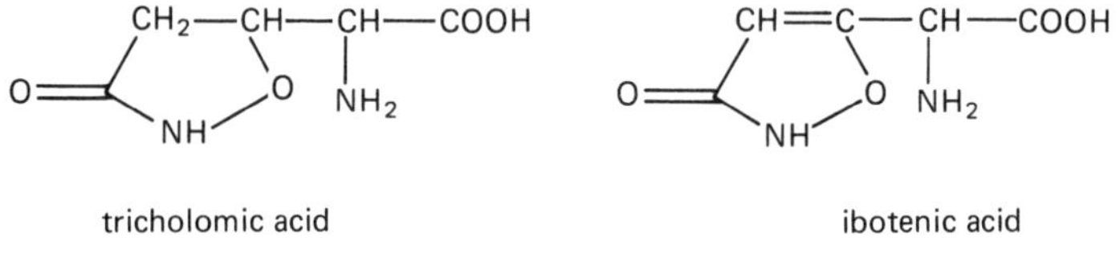

tricholomic acid

ibotenic acid

Figure 5.9   Insecticidal metabolites of basidiomycetes

Bacillus thuringiensis

This aerobic, spore-forming bacterium is pathogenic to the larvae of lepidoptera, and affords a practical means for their biological control. The larvae become paralyzed within an hour of ingesting a sporing culture of this organism, whereas vegetative cells from a young culture have no activity at all.

Microscopic examination of a sporing culture reveals a single large crystal in each rod-shaped cell, at the opposite end to the developing spore. This crystal is composed entirely of an alkali soluble protein, which dissolves in the alkaline juices of the caterpillar's digestive cavity. The crystalline protein is, in fact, a protoxin which is hydrolyzed to a smaller toxic molecule by proteases in the gastric juices. In one strain of *Bacillus thuringiensis*, 32% of the total biomass of a young sporulated culture is composed of this crystalline protein which shows such a high, specific, toxicity to lepidoptera larvae at dose levels which are comparable with those for DDT (Table 5.16).

Table 5.16 Toxicity of the protein crystals of *Bacillus thuringiensis*

Strain	Form of protein	Test insect	LD_{50} (μg/g larva)
Sotto	Digested	*Bombyx mori*	0·07
	dissolved		0·8
Berliner	Intact crystal	*Pieris brassicae*	0·25

Man and other mammals are unaffected, probably because the acid of the stomach or the action of pepsin degrades the protoxin to a non-toxic form. Useful insects, such as bees, are resistant to low doses; thus sporulated cultures or purified preparations of the protein crystals obtained by CCl_4 water partition (see Chapter 3) can be used as a selective insecticide.

Production on an industrial scale is relatively simple because *Bacillus thuringiensis* is amylolytic and grows on a mineral salts medium containing starch.

Cultures may be grown on semi-solid media based on bran, or in submerged culture, the final product being dried by vacuum spray drying.

Synthesis of the crystalline protein and spore formation are related. Electron microscope studies show that the crystal develops near the exosporium, most probably as a result of excess production of spore wall protein: the crystalline protein and the spore wall proteins are antigenically related. It is thought that a lack of control in spore wall biosynthesis leads to the formation of the crystal.

Crude preparations of the protein protoxin contain other toxic metabolites of smaller molecular weight. The most interesting, a water-soluble, heat-stable β-exotoxin, is a basic compound with a molecular weight of about 900 and is, essentially, an adenine nucleotide with an unusual sugar derivative called allomucic acid (Figure 5.10). It is active against diptera (flies and mosquitos), coleoptera (beetles), orthoptera (termites) as well as lepidoptera (butterflies and moths).

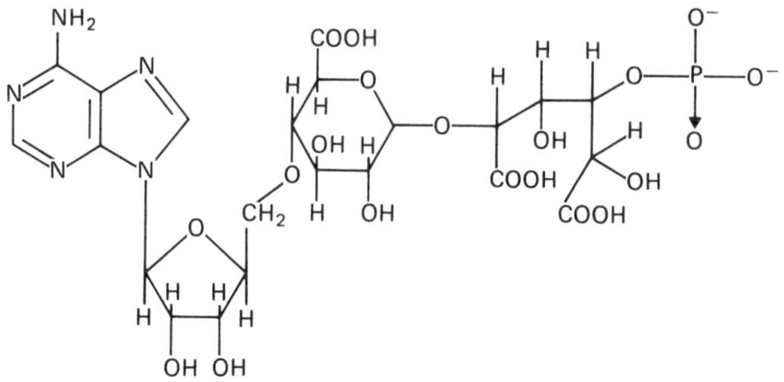

Figure 5.10   Structure of β-exotoxin of *Bacillus thuringiensis*

For adult cattle flies the exotoxin is lethal only at fairly high doses; however, a lower dose affects pupal development, and the resultant adults are incapable of flight. For controlling those cattle flies which lay their eggs in faeces, introduction of the exotoxin into the animals' feed has been investigated: administration of 6 g/day of a crude *Bacillus thuringiensis* culture preparation yields sufficient residual exotoxin in the faeces to preclude emergence of flight-capable adults.

At the molecular level, it has been shown that this toxin inhibits DNA-dependent RNA polymerase in *Escherichia coli*. It could well be that the mode of action in insects is similar, since they are sensitive to the toxin mainly during their pupal stages, when RNA synthesis is maximal. In fact, the sensitivity of different families of insects varies considerably. The lepidoptera and herbivorous hymenoptera, for example, are only sensitive to the ingestion of spores giving rise to a toxaemia, or even a septicaemia, in the post-larval stages. Several hundred tons of *Bacillus thuringiensis* preparations are manufactured in the United States and there are commercial preparations, sold under the name of Bactospeine, in France.

Bacillus popilliae

Larvae of the Japanese beetle (*Popillia japonica*) are susceptible to a disease caused by an aerobic spore forming rod-shaped bacterium (*Bacillus popilliae*). The ingested spore of *Bacillus popilliae* germinates in the digestive cavity of the larva and vegetative cells invade the body, multiplying extensively in the haemolymph. Infected larvae develop a milky appearance before death. The pathogenesis of the infection is however not well understood. Apparently, toxins are liberated into culture fluids, for filtrates are lethal when injected in small quantities. It is also possible that, whilst multiplying in the haemolymph, the bacteria utilize nutrients at the expense of the host.

Under field conditions, some 10^7–10^9 spores per gram of soil are required to bring about infection of 50% of a natural population of larvae. However, *in vitro* production of sufficient spores is difficult, except with certain strains of low virulence which apparently germinate only with difficulty in the haemolymph. A powder containing 10^8 virulent spores per gram can be prepared from infected larvae which yield, after 7–12 days, 5×10^9 spores per individual (or 5×10^{10} spores per ml of haemolymph), but use of such a product for spreading on soil is not economically feasible. Further research is required to produce spores of virulent strains in deep fermenters before this organism can be used as a practical biological control agent.

Nematodes

Those nematodes which act as vectors of insect pathogens can sometimes be used in the biological control of insects. An example is the symbiotic association between *Achromobacter nematophilus* and *Neoaplectana carpocapsae* (the so called nematode DD 136) which harbours the bacterium in its intestinal cavity. The bacteria are released from the anus when the nematode penetrates an insect host such as the larva of the apple moth (*Carpocapsa pomonella*). Nematode DD 136 can be cultured, with its bacterial partner, on artificial media and a technician can inoculate about 200 Petri dishes per day, recovering 10^8 nematodes in a week. At low temperatures, these remain viable for 4–6 weeks. Even so, environmental conditions, such as humidity and temperature, play an important role in maintaining the association between bacterium and nematode to ensure a population pathogenic to the insect.

Fungal pesticides

There are many fungi, some of which penetrate the integument of their host without needing to be ingested. Their pathogenicity is frequently dependent on environmental factors such as temperature and humidity: certain combinations of such conditions enable the spores to germinate and the mycelium to invade the body of the host insect and liberate its toxins. One of the best known is *Beauveria bassiana* which is particularly useful in the destruction of moth larvae, although it has also been the cause of a destructive disease of the silkworm (*Bombyx mori*).

Conidiospores of *Beauveria* spp. may be harvested from cultures on semi-solid media, giving a powder containing 5×10^9 spores/g. Unfortunately conidia cannot be obtained in submerged culture, although the much more fragile blastospores are produced. Similar results are obtained using *Metarrhizium anisopliae*.

These fungi act in some respects like contact insecticides, having a wide spectrum of activity, but slow in effect. Since there have been indications of human allergy to the spores, they are not, in fact, produced commercially.

Viruses pathogenic to insects

A number of viruses are pathogenic to insects. The most interesting occur in the form of protein inclusions, the polyhedroses, in which several virus particles are contained in each polyhedral protein crystal, and the granuloses, in which each crystal contains a single virus particle. When ingested by an insect larva, the protein crystal dissolves in the digestive cavity and the released virus particles enter host cells, multiplying in either cytoplasm or nucleus. When the infected larva dies, enormous numbers of new virus particles are liberated, each protected by a protein coat which may represent about 30% of the dry weight and confers a high degree of resistance to adverse environmental conditions and lengthy infectivity. The only means currently available for producing these viruses entails breeding and infecting large numbers of larvae in the laboratory and are not economic.

The polyhedrose viruses of the cabbage worm (*Trichoplusia ni*) and the cotton boll worm (*Heliothis zea*) have been most intensively studied. Virus suspensions, together with a wetting agent, are applied to the foliage of the infected plants when the larvae are very young and death occurs within 1–7 days (Table 5.17).

Table 5.17 Effect of viral treatment of a field of maize on the destruction of the larvae of the insect *Heliothis zea* (3 ml Triton X-100 were added to each 16 gallons of suspension used per acre)

Test	Date of application	Polyhedra per gallon*	Reduction in population	Reduction in damage
1	27th April	3.7×10^{10}	64%	54%
2	1st May	1.9×10^{10}	68%	48%
3	1st May	3.7×10^{10}	86%	83%

* 1 gal = 3·785 litres: 1 acre = 4047 m^2.

Microbiological control agents of the kind described here have important potential advantages over chemical pesticides. They can be chosen so as to have a high degree of specificity for the target host, with little risk to man, domestic animals or even arthropods beneficial to plant crops. Unlike certain chemical pesticides, they do not accumulate in the environment or undergo concentration by entering food chains; in the absence of the target host they frequently disappear completely. They are, to some extent, self-propagating in a target-species population of sufficient density. However, most such agents are in the experimental stages of study: those for which economic methods of culture or propagation can be developed may well represent an important future area in industrial microbiology.

Gibberellins

The gibberellins are a group of metabolites with complex chemical structures based on the tetracarbocyclic gibbane nucleus (Figure 5.11). They are produced by plants, in which it is believed they play an important role as hormones, but they are also synthesized by a number of moulds. Although they can be isolated from such plant material as immature bean seeds the gibberellins are manufactured using the mould *Fusarium moniliforme* which is the imperfect state of *Gibberella fujikuroi*, a plant pathogen responsible for a disease of young rice plants. The disease appears as an elongation of the internodes leading to a collapse of the plant at an early stage of growth. Under the right conditions some strains of *Gibberella fujikuroi* produce only one gibberellin known as gibberellic acid or gibberellin A_3 (Figure 5.11), which, in malting, is sometimes applied to the germinating barley because it increases the yield of amylases.

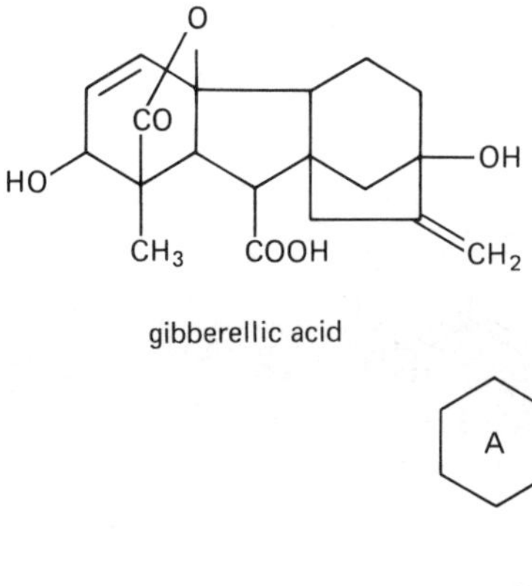

gibberellic acid

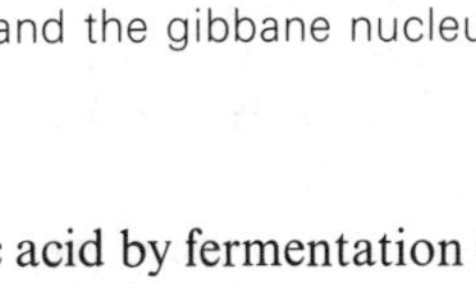

gibbane nucleus

Figure 5.11   Gibberellic acid and the gibbane nucleus

Media for the production of gibberellic acid by fermentation should be low in nitrogen and should contain a mixture of carbohydrates; the presence of glycerol particularly increases the yield. Gibberellic acid has received much experimental attention because of its potential applications in agriculture (to increase crop yields) and horticulture (to induce flowering of plants independently of photoperiodism); its practical use has so far been in a few specific applications such as obtaining seedless grapes.

Alkaloids of microbial origin

The first microbial alkaloids to be recognized and studied were those of
Claviceps purpurea—the agent causing ergot of rye. These alkaloids can be
isolated from the sclerotia formed after infection of the ovaries of the plant by
Claviceps ascospores or conidia; they all contain the tetracyclic nucleus of
ergoline (Figure 5.12) and can be classified into three groups, clavine,
isolysergic acid and lysergic acid (Figure 5.13).

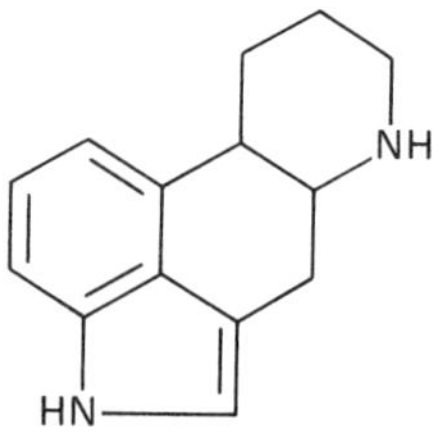

Figure 5.12   The tetracyclic structure of ergoline

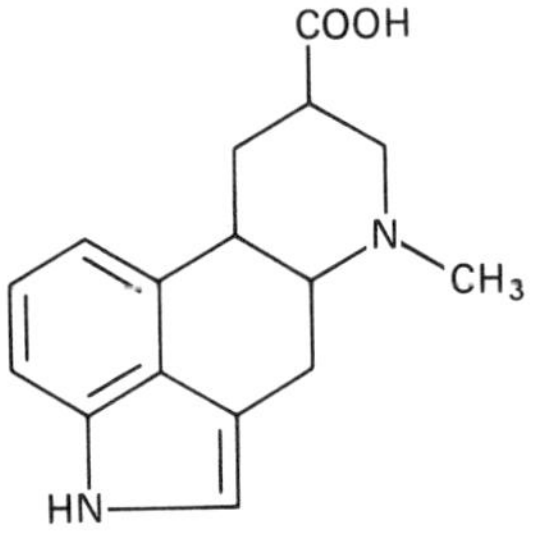

Figure 5.13   Lysergic acid

These compounds all have powerful pharmacological actions affecting the
sympathetic nervous system. Probably the best known, and certainly the most
infamous, is the diethylamide of lysergic acid (LSD), a powerful and dangerous
hallucinogen; it has limited therapeutic applications but its psychomimetic
effects have led to misuse.

Several strains of *Claviceps purpurea* and of *Claviceps paspali*, a related
fungus producing sclerotia on the tropical grass *Paspalum*, can be grown as
submerged culture in deep fermenters. High yielding strains do not readily
produce conidia: it is therefore necessary to use a mycelial inoculum which can
be preserved, if required, at low temperatures in a medium containing 20 %
glycerol. High yields of alkaloids (<2 g/l) are obtained using media containing
high concentrations of sugars (300 g/l of sucrose). Fermentation takes up to 10

days and the frequently complex mixture of alkaloids is extracted from the mycelium with chloroform after pH adjustment to 8–10.

Yields of various alkaloids from a fungal parasite of *Pennisetum typhoideum*, the bullrush millet, are given in Table 5.18.

Table 5.18 Yields of alkaloids from a fungal parasite of *Pennisetum typhoideum* grown in a liquid medium* for 10 days at 24°C

Alkaloids extracted	Yield (g/100 l medium)
Elymoclavine	69·9
Agroclavine	69·4
Setoclavine	2·1
Isosetoclavine	1·9
Penniclavine	1·9
Isopenniclavine	0·1
Chanoclavine	0·6

* Sucrose, 100 g; ammonium succinate, 10 g; calcium nitrate, 1 g; monopotassium phosphate, 0·25 g; magnesium sulphate, 0·25 g; potassium chloride, 0·125 g; ferrous sulphate, 8·34 mg; zinc sulphate, 3·44 mg; water, 1 l.

One particular strain, C_1M, has a derepressed tryptophan synthetase system and forms a number of biologically active derivatives of tryptophan from the corresponding substituted indoles when these are present in the medium. Some of these derivatives have been shown to inhibit viral replication.

Finally, alkaloids are susceptible to microbial transformations analogous to those used in the manufacture of steroids (Table 5.19).

5.3 Enzymes

The metabolic products so far considered (primary and secondary metabolites) are the results of enzymic activities of microorganisms. In some industrial fermentations, the enzyme itself is the desired product. Enzymes produced by fermentation are mostly extracellular ones but a few important examples (invertase; asparaginase; uric oxidase) are endocellular and more difficult to manufacture because of the necessity to disrupt the cells before extraction and purification (see Figure 3.17).

Table 5.19   Microbial transformations of alkaloids

1. Hydroxylation

   1.1. $>CH_2 \longrightarrow >CH-OH$

   1.2. $-CH< \longrightarrow -C<-OH$

2. Reduction

   2.1. $>C=O \longrightarrow >CH-OH$

   2.2. $-CH=CH- \longrightarrow -CH_2-CH_2-$

3. Hydrolytic

Composition of the culture medium

Many of the useful microbial enzymes are hydrolases. These are frequently
'induced' enzymes, that is to say they are produced by the microorganism only
when it grows on a medium containing the enzyme's substrate (e.g. β-
galactosidase may be produced only when the medium contains lactose). Less
frequently, there is an 'end-product effect', by which formation of the enzyme is
stimulated by the presence of the end-product of hydrolysis (e.g. cellulase
production may be stimulated by the presence of cellobiose). Enzyme
biosynthesis can be inhibited by catabolic repression when the medium
contains a readily available, easily metabolized energy source, such as glucose
or some of the possible inducers of polysaccharases. For example, 0.5%
cellobiose may inhibit cellulase biosynthesis, by catabolic repression, despite
cellobiose being an inducer of the enzyme. One method of obtaining high
enzyme yield is to maintain a constant low concentration of the inducer (e.g.
0.05% cellobiose) by continuously feeding it into the medium. Another is to
add the inducer to the medium in the form of a derivative which is slowly
broken down, maintaining a constant but very low concentration of inducer.
An example of the latter approach is the use of hexose esters, which may
increase yields of certain enzymes by overcoming catabolic repression, so long
as the microorganism producing the enzyme also synthesizes the appropriate
esterase (Table 5.20). The black yeast, *Aureobasidium pullulans*, which produces a monopalmitate esterase, will synthesize 100 times more invertase than
brewing yeasts: furthermore, *Aureobasidium* invertase, unlike that of
Saccharomyces, is extracellular.

Table 5.20 Effect of various inducers on
yields of selected enzymes*

Enzyme	Culture	Inducers	Yield in international units
Cellulase	*Trichoderma viride*	Cellulose	22·5
		Cellobiose	0·2
		Cellobiose dipalmitate	4·8
Dextranase	*Penicillium funiculosum*	Dextrans	1080
		Isomaltose	2
		Isomaltose dipalmitate	1098
Invertase	*Aureobasidium pullulans*	Sucrose	1·3
		Sucrose monopalmitate	108

* Reese *et al.*, (1969)

It may be possible to select the most slowly metabolized of a number of inducers, for example, xylidine gives high laccase yields by *Coriolus versicolor* (= *Polyporus versicolor*). Again, in the production of tyrosinase by *Neurospora crassa*, the less readily metabolized D-isomers, such as D-tyrosine and D-phenylalanine, are better inducers than the corresponding L-isomers.

When a compound acting as a repressor is also an indispensable metabolite for the microorganism producing the enzyme, (e.g. a purine base or an amino acid) then some slowly used derivative or precursor of the compound may be added to the medium. Thus, adenosine is an inducer of purine nucleosidase but at 0·5% concentration it acts as a repressor. However, RNA can be used by *Aspergillus ambiguus* as a source of adenosine for nucleosidase production.

To elucidate the nutritional requirements for the production of a required enzyme, defined or synthetic media may be used. Once it is understood what is indispensable, a search can be made for cheaper materials which will give the same result. However, since in the production of pure enzymes media costs represent only about 1% of the total, attention may better be directed towards devising media which make subsequent recovery of the enzyme easier. Empirical determination of the optimum concentration of each constituent involves a considerable amount of work. Because small peptides often play an important role as inducer, a complex nitrogen source is frequently required. It is only rarely that the function of a particular constituent has been precisely determined, but it is known that arginyl proline stimulates the invertase biosynthesis by *Saccharomyces fragilis;* biosynthesis of α-amylase requires the presence of calcium and magnesium; the alkaline protease of *Bacillus subtilis* is produced in optimum concentrations at neutral pH. These few examples indicate the fragmentary nature of information available.

Different enzymes produced by a single organism do not always appear at the same stage of growth. Thus *Apergillus niger* may produce both α-amylase and maltase: the yield of α-amylase is highest after 48–72 h incubation at 35°C; thereafter, maltase increases in concentration as α-amylase decreases. After sufficiently lengthy incubation, maltase can be obtained, free of α-amylase.

Production of a particular enzyme is sometimes restricted to a very specific period of growth, as in the case of L-asparaginase production by strains of *Escherichia coli*. Unlike other enzymes, L-asparaginase synthesis is not repressed by the presence of glucose, but the conditions of production are very specific. When the organism is grown aerobically, yield of biomass is high but little enzyme is produced. Anaerobically there is a relatively poor yield of biomass but cells are rich in asparaginase. In such cases it may be necessary to employ a two stage process, the first to build up a good yield of biomass, the second to allow the cells to synthesize the required enzyme. Good yields of L-asparaginase are obtained when the culture is incubated aerobically for 8 hours, bringing it to the end of the exponential phase, and then maintained anaerobically for no more than 2 hours: the cells are then harvested and the enzyme isolated.

The addition of surfactants to the culture medium increases yields of certain enzymes. Non-ionic detergents such as Tween 80 are frequently used for others may be toxic (Table 5.21). Occasionally Triton (Rohm and Haas) produces better results.

Table 5.21   Effect of 0·1 % Tween 80 on the yield of various enzymes

Enzyme	Producing microorganism	Yield in medium + Tween 80 / Yield in medium (ratio)
Cellulase	*Trichoderma viride*	20
Invertase	*Aureobasidium pullulans*	16
β-1·3 glucanase		8
β-glucosidase		4
Xylanase	Numerous fungi	4
Amylase		5
Nucleosidase		6
Esterase		
Dextranase	*Penicillium funiculosum*	2
Pullulanase	*Aerobacter aerogenes*	1·5

The mode of action of these surfactants is not understood. It is possible that they modify cell permeability, accelerating release of the enzyme into the medium: at the same time, because of various feed back control mechanisms operating to maintain a constant concentration of enzyme within the cell, the rate of synthesis is increased.

Uses and sources of microbial enzymes

Amylases

Strains of *Bacillus* and *Aspergillus* produce a mixture of three enzymes in varying proportions:

1. an α-amylase which acts on starch producing dextrins and maltose. Because starch is liquified and solutions reduced in viscosity, this enzyme is sometimes referred to as liquifying enzyme;
2. an amyloglycosidase which produces glucose directly from starch;
3. a maltase which hydrolyzes maltose to glucose.

Fungal amylases (manufactured using *Aspergillus niger* and *Aspergillus oryzae*) are used to hydrolyze starch to fermentable sugars. In the brewing industry these amylases are used to control starch hazes, and also allow the production of beers free from dextrins, the so called low calorie beers. Partial pre-digestion of certain foods with fungal amylases makes them suitable for very young children. Amylases are used to prepare syrups from maize and wheat starch and to eliminate starch from fruit juices to produce drinks with low turbidity. In bread making, the addition of a certain amount of amylase to flour results in a more controlled production of carbon dioxide by yeast cells during raising, giving a lighter loaf. Sizes and other coating materials used in the manufacture of paper are produced from starch by treatment with fungal amylases. In the confectionery industry they are used to produce chocolate syrups from cocoa.

Bacterial amylases are much more stable to increased temperatures. One use is in the textile industry where, during weaving, the warp is subjected to considerable strain which may cause it to break; thread for the warp may be pre-strengthened by passing it through a starch preparation; when weaving is complete, the cloth is washed in a dilute solution of bacterial amylases at 50°C to remove the starch. The same enzymes have also been widely incorporated into 'biological detergents' for domestic laundry purposes: some early formulations were believed to have induced allergies in a few users, due to the crude nature of the enzyme preparations used.

β-galactosidase (lactase)

In the manufacture of ice cream, lactose crystallization may impart a disagreeable texture to the product, as though it contained fine hair. This can be prevented by adding β-galactosidase, obtained from lactose-fermenting yeasts such as *Saccharomyces fragilis* or *Candida pseudotropicalis*. These yeasts grow well in milk whey and cheese wastes, to which it is necessary to add some nitrogen compounds (0·06% ammonia); a low aeration rate is adequate. Cheap final preparations can be obtained by vacuum drying, which destroys any zymase present, this being the complex mixture of enzymes and coenzymes which brings about the fermentation of glucose, a process which would be undesirable in a product such as ice-cream.

Invertase

This enzyme is readily obtained from brewing yeasts (*Saccharomyces cerevisiae* and *Saccharomyces carlsbergensis*) grown in a sucrose based medium; selected high-invertase strains are used. The enzyme is endocellular and is liberated from the cells by autolysis in the presence of toluene (cf. pp 101–102). After filtration from cell debris, the enzyme is precipitated by adding organic solvents. Alternatively, the whole autolyzed preparation may be dried, invertase being quite stable to such treatment. Invertase hydrolyzes sucrose to a mixture of glucose and fructose, both of which are more soluble. Sucrose-invertase mixtures are therefore used in the manufacture of chocolates or other sweets with liquid centres.

Glucose isomerase

Fructose is much sweeter than glucose and very sweet syrups can thus be prepared from glucose solutions by treatment with glucose isomerase. The enzyme has been isolated from a number of microorganisms, including strains of *Streptomyces* and *Actinoplanes*, and is now manufactured on a large scale using *Bacillus coagulans* (Novo) and *Arthrobacter* sp. (ICI). The enzyme may be used in an immobilized form in a plug-flow column reactor; concentrated solutions of glucose are rapidly converted to an equilibrium mixture containing approximately equal amounts of glucose and fructose. Such a mixture is as sweet as an equivalent weight of sucrose. The concentrated glucose solutions are themselves readily obtainable by the action of fungal amylases on maize starch.

Proteases

This term is usually employed to describe complex mixtures of true proteinases and peptidases. The latter, being endocellular, are present only in low concentrations in the medium at the end of growth. Proteases may be produced by bacteria (e.g. *Bacillus subtilis* and *Bacillus licheniformis*) or by fungi (*Aspergillus niger*, *Aspergillus oryzae*). They require some care in preparation because they are relatively unstable and tend to lose their activity during dehydration.

Bacterial proteases are widely used. In the tanning of leather they now replace the trypsin treatment which Rohm, in 1905, introduced as an alternative to the nauseating traditional process involving fermentation with dog faeces and urine. Protease treatment of hides causes a slight hydrolysis giving softer texture. Proteases from *Streptomyces fradiae* cultures are used for graining and dewooling the skins.

Bacterial proteases are also employed in the textile industry. They are used in silk production to dissolve a gummy substance ('sericine') which surrounds the natural protein fibre. During weaving silk and other fibres, including man-made ones, may require lubrication and strengthening with gelatin, casein or other proteinaceous substances. These substances have to be removed from the woven material during the process of desizing, and the use of bacterial proteases is common.

Proteases have also been used to hydrolyze the gelatin base of used photographic films in order to recover their silver content: the silver content of used radiographic films is particularly high. Proteases have been increasingly used in the so called biological detergents, intended for removing protein-aceous stains such as blood, mucous, coffee and chocolate from clothing and other fabrics. Alkaline proteases, produced by many strains of *Bacillus* and *Streptomyces* have been mainly used because of their greater stability at high temperatures. Alkaline proteases tolerate high pH and temperatures, and are active in the presence of such detergent additives as sodium perborate, sodium tripolyphosphate and sodium alkylbenzene sulphonate (Table 5.23).

Table 5.22 Composition of the synthetic base of a 'biological detergent'

Product	Proportion (%)
Sodium alkylbenzene sulphonate	25
Alkylphenol ethoxylate	3
Savon	3
Carboxymethyl cellulose	1
Sodium ethylene diaminotetracetate	0·2
Sodium polyphosphate	38
Sodium perborate	25
Sodium metasilicate	1
Sodium sulphate and water	to 100

The molecular weight of alkaline proteases is 20,000 to 30,000 and their active site contains serine and histidine. They lack disulphide bridges, and are thus resistant to oxidation by perborate.

These impure microbial enzymes have no demonstrable toxicity to man or animals, but instances of hypersensitivity have been reported; their general use has been criticized on this ground. On the other hand, it can be argued that they are very readily biodegradable.

Table 5.23 Stability of proteases to domestic detergent components and conditions

| | Residual activity (%) in the presence of: | | | | |
Enzyme	Tripolyphosphate 0·1%*	Perborate 0·1%*	pH12	60°C	Alkylbenzene sulphonate 0·05%*
Papain	100	0	0	100	60
Trypsin	11	48	15	30	0
Alkalase					
(*B. licheniformis*)	44	90	60	100	50
SP.72 (*Bacillus* sp.)	89	90	90	100	80

* Residual activity measured after 30 minutes at 50°C and pH 9·0

Microbial rennins

Coagulation of milk prior to cheese manufacture is usually by rennet extracted from the abomasum of young, milk-fed calves.

Calf rennet acts on casein K, (one of the three milk caseins α, β and K) responsible for maintaining the stable suspension of micelles, by breaking its phenylalanine-methionine bonds; the hydrolysis products are paracasein K and a soluble glycopeptide. In the presence of calcium ions, precipitation of caseins α and β and of paracasein K occurs and a soft clot forms. Microbial rennet must have similar properties without undue further proteolytic activity which would ultimately redissolve the clot. The enzymes of many bacteria and fungi have been examined for their suitability in cheese making. If proteolytic activity is too high, short peptides are produced giving rise to a bitter taste in the finished product. The microbial rennets commercially produced are of fungal origin (*Endothia parasitica*, *Mucor pusillus* and *Mucor miebei*) and cheeses of good quality can be obtained using them.

Pectinases

Pectinases prepared from strains of *Aspergillus* and *Penicillium* are commercially available. Beet sugar molasses contains pectins and provides an excellent culture medium for their production. After growth is complete, the mycelium is dried and pulverized and a mixture of pectin esterase, polygalacturonase, polymethylgalacturonase and pectin transeliminase extracted with cold water. This crude enzyme mixture is used to clarify fruit juices, which can then be concentrated and refrigerated without solidifying. Pectinases are also used to eliminate the mucilage surrounding freshly picked coffee beans, a process formerly involving a natural, and usually uncontrolled, fermentation.

Lipases

Lipases produced from a culture of *Rhizopus* have been used for degreasing wool.

Glucose oxidase and catalase

These enzymes are manufactured using such fungi as *Aspergillus niger* and *Penicillium notatum*. The enzyme mixture is used in the food industry to eliminate glucose from certain food products such as dried egg white, to prevent discoloration due to chemical reactions between glucose and free amino acids referred to generally as browning reactions (Figure 5.14).

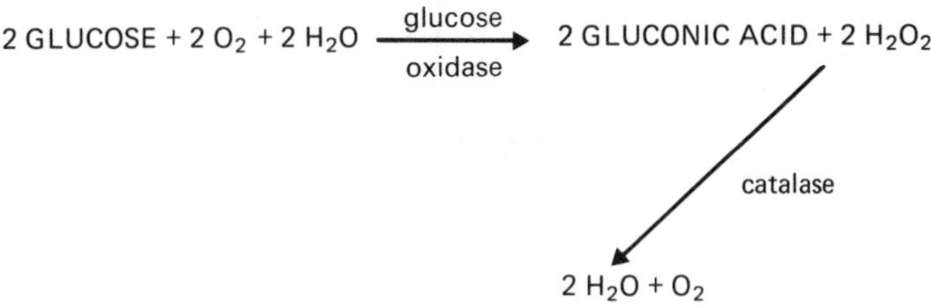

Figure 5.14   Action of a mixture of glucose oxidase and catalase in the removal of glucose

   Glucose oxidase is also used in the specific assay of the glucose component of blood sugars.

Diacetyl reductase

This enzyme, extracted from cells of *Enterobacter aerogenes*, is used for the removal of diacetyl from beer and fruit juices, to which excessive quantities can impart off-flavours. The enzyme acts in the presence of NADH$_2$ which is provided by adding yeast cells. It is sometimes employed in the brewing of light beers by continuous fermentation.

Naringinase

This enzyme, produced by certain strains of *Aspergillus niger*, can be used to remove the bitter substance, naringin, present in the juice of some types of orange. The enzyme preparation must also contain flavonoid glucosidase activity (Figure 5.15); if the mould is grown on a lactose based medium, an enzyme is obtained which is not inhibited by the presence of glucose in the orange juices.

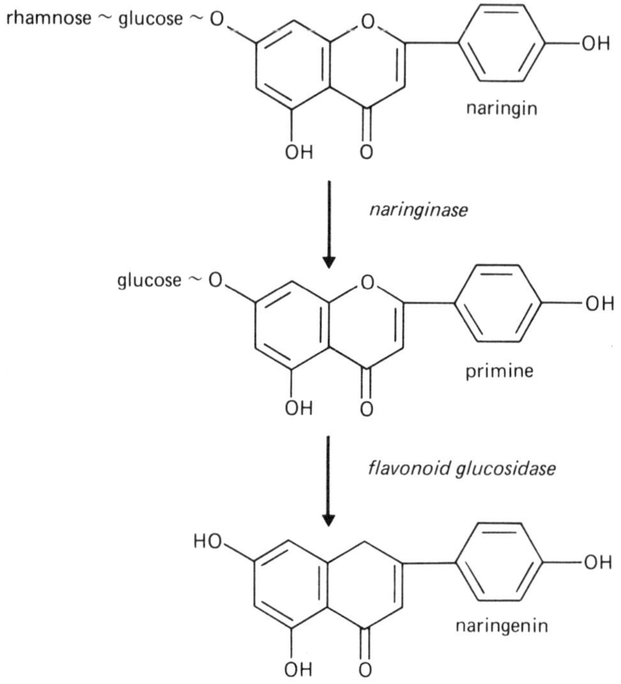

Figure 5.15   Hydrolysis of naringin

Medical applications of microbial enzymes

The medical applications of enzymes are numerous. Microbial hydrolases can be used to supplement a deficiency in the production of digestive enzymes. Thus acid resistant amylases, produced by *Aspergillus oryzae* and *Aspergillus niger*, α-amylase from *Bacillus subtilis* and cellulases from *Trichoderma viride*, are used as digestive aids; fungal proteases may also be used for the treatment of alimentary dyspepsia because they are also resistant to acid; lipases produced by *Aspergillus oryzae* or *Candida lipolytica* are sometimes used to compensate for a deficiency in pancreatic lipase.

A fungal lactase is added to preparations of milk products to reduce the reaction due to intolerance to milk which frequently occurs in populations of developing communities. Some children, particularly in developing communities, are intolerant to milk because of a deficiency in β-galactosidase: lactose is not hydrolyzed and absorbed but instead ferments, with gas production, in the large intestine. The addition of fungal lactase to milk reduces this reaction.

Early in dental caries, before decay begins, bacterial dextrans are deposited on the tooth surface. Attempts have been made to inhibit this deposition using a dextranase produced by *Penicillium funiculosum.*

L-asparaginase has been used in the treatment of certain types of leukemia in which the neoplastic cells require exogenous L-asparagine being unable, unlike normal cells, to synthesize their own. This discovery arose from an early observation that mouse lymphoma could be suppressed by the injection of guinea-pig serum. It later emerged that the factor responsible for recovery was L-asparaginase, which hydrolyzes asparagine to aspartic acid and ammonia. The enzyme used for treating L-asparaginase-sensitive leukemias is obtained from a strain of *Escherichia coli,* yields being up to 1 g/l. Since effective treatment requires intravenous injection of the enzyme, all traces of the endotoxins characteristically produced by members of the Enterobacteriaceae have to be eliminated by careful purification. Because the enzyme is an antigenic protein, antibody appears during the course of treatment and effectively neutralizes later doses. Treatment can be continued by using L-asparaginase from antigenically different species such as *Erwinia carotovora* var. *carotovora.* By using a sequence of different asparaginases, treatment may be prolonged.

The use of enzymes for treatment of thrombosis (intravascular blood clot formation) has also been investigated; both prevention of clot formation and dissolution once a clot has formed are possible. Streptokinase, an enzyme produced by some β-haemolytic streptococci can be injected intravenously to activate plasminogen by splitting the bond between arginine and valine, transforming it to plasmin (fibrinolysin). Plasmin causes digestion and solubilization of the fibrin network which, together with entrapped erythrocytes, white cells and blood platelets, makes up the thrombus or clot. Encouraging results have been obtained with highly purified streptokinase in the treatment of myocardial infarctions.

High doses are required, partly because human blood contains natural antistreptokinase, and the enzyme preparation has to be freed of pyrogenic substances. Active research has continued to discover other microorganisms producing enzymes with analogous properties and *Cephalosporium* strains and *Aspergillus oryzae* B 1273 have all shown promise.

Streptodornase is another enzyme of medical usefulness produced by haemolytic streptococci. Its substrate is DNA or deoxyribonucleoproteins, which are responsible for much of the viscidness of purulent exudates. Streptokinase and streptodornase are almost always used together: mixtures can be injected into fibrinopurulent accumulations in body cavities to bring about their liquefaction and dispersal, and can also be inhaled as aerosols to cause thinning of tenacious bronchial secretions.

A pure, toxin-free, collagenase from *Clostridium histolyticum* has been used for enzymic debridement of necrotic, infected tissue in burns or certain forms of skin ulcer. This enzyme preparation has also been used for treating skin homografts before transplanting them onto areas of extensive burns. Collagenase injections have also been tried in the correction of slipped intervertebral discs, which can cause persistent pain by compressing nerve roots.

Although penicillinase production is a nuisance in the manufacture of penicillins, and in reducing the spectrum of activity of certain penicillins, it is manufactured for a number of uses. Penicillinase, obtained from *Bacillus cereus*, is used, by intramuscular injection, in the medical emergency which occasionally arises when a hypersensitive patient who has been dosed with penicillin develops an allergic or, worse, anaphylactic reaction; it causes hydrolysis of penicillin to the biologically inactive penicilloic acid. Penicillinase is also used in the specific assay of penicillins especially in the presence of other antibiotics.

A urate oxidase, prepared from a strain of *Aspergillus flavus*, has been used in the management of gout: it is claimed to aid dispersal of uric acid crystals.

Methods of using microbial enzymes

The earliest approach was to use solutions of enzymes. After a brief period of use, such solutions lose their activity and have to be discarded or subjected to expensive and difficult repurification procedures. However, many of the enzymes of the microbial cell do not operate in solutions, but are fixed on certain structures such as the cell membrane. Accordingly, there have been many studies on the possibility of using microbial enzymes immobilized or insolubilized in some way.

There are five principal methods for insolubilization of enzymes:

1. Adsorption onto an inert support substance such as a resin, clay, sintered glass or glass beads: the latter can easily be used as a column.
2. Incorporation in the mesh of a polyacrylamide or other gel; the enzyme is immobilized, whilst the substrate and product of the reaction can diffuse through the gel spaces.
3. Covalent linkage of the enzyme with a polymer using functional groups well separated from the active site of the enzyme. An example is the reaction of free amino groups of an enzyme with the hydrazide of carboxymethyl cellulose (Figure 5.16).
4. Covalent bonding of the enzyme with itself using bifunctional compounds such as glutaraldehyde.
5. Retention of the enzyme, which may be soluble, by means of an ultrafiltration membrane.

In many cases, immobilized enzymes prove to be stable and fairly resistant to changes in temperature and pH of the medium. They can then be used in automated continuous processes. Sometimes their preparation is difficult,

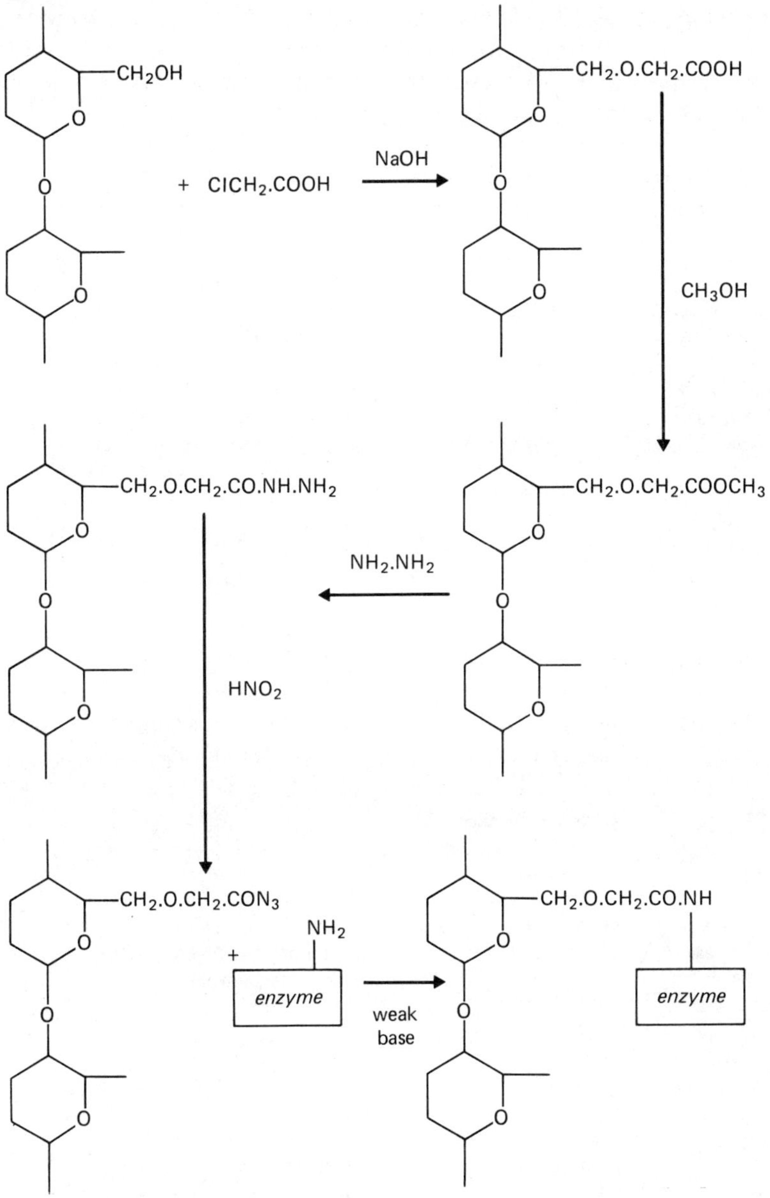

Figure 5.16 Preparation of bound enzymes using carboxymethyl cellulose (CMC)

their durability limited and performance mediocre due to steric hindrance.

A number of practical applications have been forecast for the near future in medicine and the pharmaceutical industry, bioconversions, synthesis of polymers, inversion of sucrose, saccharification of starch, enzyme electrodes and antipollution measures. Table 5.24 gives some examples.

Table 5.24 Examples of microbial enzymes bound to insoluble supports

Support	Enzyme	Bonding
Cellulose	β-galactosidase	2,4,6, trichlorotriazine
Carboxymethyl cellulose	Asparaginase	Ionic
Aminoethyl cellulose	Aldolase	Glutaraldehyde
Starch gel	Urease	Inclusion
Agar	Hexokinase	Cyanogen bromide
Polyaminostyrene	Invertase	Diazonium salt
Polyurethane	Glucose oxidase	p-amino hippuric acid—protein, N,N dicyclohexylcarbodiimide
Nylon	Urease	Glutaraldehyde
Sintered glass	Ribonuclease	Adsorption
Hydroxyapatite	Glucose oxidase	Derivative of isothiocyanoalkyl silane

Enzyme columns may be used continuously and an aminoacylase bound to DEAE Sephadex has been shown to be stable for up to twenty days. Columns of bound enzymes are already used industrially to obtain L-amino acids from racemic mixtures. It has been suggested that columns containing bound phenol

Table 5.25 Industrial uses of bound enzymes

Enzyme	Support	Uses
Amino acid acylase	DEAE cellulose	To obtain L-amino acids
Penicillin acylase	Triazinylcellulose	Manufacture of 6-APA (to produce semisynthetic penicillins)
Amyloglucosidase	DEAE cellulose, glass	Production of glucose from starch
Glucose isomerase	Polyacrylamide, glass	Production of fructose from glucose
Invertase	DEAE cellulose	Hydrolysis of sucrose
α-amylase	Glass	Paper manufacture (elimination of starch from waste waters)
α-galactosidase	Glass	Elimination of the factor responsible for intestinal upset from soya flour
β-galactosidase	Cellulose, glass	Hydrolysis of lactose in milk or serum
Proteases	Glass	Clarification of beer

oxidases might be used to treat chemically polluted water. Table 5.25 lists some industrial uses of insoluble enzymes.

Potential medical applications, largely in the experimental stage, include the use of micro-encapsulated enzymes implanted in the body in nylon or collodion sacs. These might be used to eliminate toxic metabolites, as for example in phenylketonurea; if urease were the bound enzyme, it might be possible to remedy some degree of kidney dysfunction. The possibility of implanting enzymes such as L-asparaginase, in a bound form, has been studied in experimental animals.

Bibliography

AKIYAMA, S., SUZUKI, T., SUMINO, Y., NAKAO, Y. & FUKUDA, H., (1972), 'Production of citric acid from n-paraffins by fluoroacetate sensitive mutant strains of *Candida lipolytica*', *4th Intern. Ferment. Symp.*, Kyoto.

BURGES, H. D. & HUSSEY, N. W., (1971), *Microbial Control of Insects and Mites*, Academic Press, New York.

DEMAIN, A. L., (1971), 'Microbial production of food additives' in 'Microbes and Biological Productivity', *Symp. Soc. Gen. Microbiol.*, **21**, 77–101.

FAITH, W. T., NEUBECK, C. E. & REESE, E. T., (1971), 'Production and application of enzymes', *Adv. Biochem. Engin.*, **1**, 77–112.

GOUGH, D. A. & ANDRADE, J. D., (1973), 'Enzyme electrodes', *Science*, **180**, 380–384.

GUTCHO, S. J., (1974), *Microbial Enzyme Production*, Noyes Data Corporation, New Jersey.

HUANG, H. T. & SHAPIRO, M., (1971), 'Insecticidal activity of microbial metabolites', *Progr. Ind. Microbiol.*, **9**, 79–112.

HUGHES, D. H., (1971), 'Enzymes as additives in detergents', *Dev. Ind. Microbiol.*, **12**, 11–56.

JEFFERYS, E. C., (1970), 'The giberellin fermentation', *Adv. Appl. Microbiol.*, **13**, 283–310.

KIKUCHI, M., SUZUKI, M., DOI, M. & NAKAO, Y., (1972), 'Microbial production of L-glutamic acid from n-paraffins by glycerol autotrophs', *4th Intern. Feremtn. Symp.*, Kyoto.

KINOSHITA, S., (1959), 'The production of amino acids by fermentation processes', *Adv. Appl. Microbiol.*, **1**, 201–212.

KULHANEK, M., (1970), 'Fermentation process employed in vitamin C synthesis', *Adv. Appl. Microbiol.*, **12**, 11–28.

NOYES, R., (1969), *Monosodium Glutamate and Glutamic Acid*, Noyes Develop. Corp., Park Ridge, USA.

NOYES, R., (1969), *Citric Acid Production Processes*, Noyes Develop. Corp., Park Ridge, USA.

ONISHI, H. & SUZUKI, J., (1969), 'Microbial production of xylitol from glucose', *Appl. Microbiol.*, **19**, 1031.

PEPPLER, H. J., (1967), *Microbial Technology*, Reinhold Publishing Corporation.

PERLMAN, D., (1959), 'Microbial synthesis of cobamides', *Adv. Appl. Microbiol.*, **1**, 87–112.

PERLMAN, D. & PERUZZOTTI, G. D., (1970), 'Microbial metabolites as potentially useful pharmacologically active agents', *Adv. Appl. Microbiol.*, **12**, 277–288.

REHAZEK, Z. *et al.*, (1971), 'Correlation of certain alterations in metabolic activity with alkaloid production by submerged *Claviceps*', *Appl. Microbiol.*, **22**, 949.

ROGOFF, M. H., (1966), 'Crystal forming bacteria as insect pathogens', *Adv. Appl. Microbiol.*, **8**, 291–312.

SARDINAS, J. L., (1972), 'Microbial rennets', *Adv. Appl. Microbiol.*, **15**, 39–66.
SIZER, J. W., (1964), 'Enzymes and their applications', *Adv. Appl. Microbiol.*, **6**, 207–226.
SIZER, J. W., (1972), 'Medical applications of microbial enzymes', *Adv. Appl. Microbiol.*, **15**, 1–11.
SMILEY, K. L. & STRANDBERG, G. W., (1972), 'Immobilized enzymes', *Adv. Appl. Microbiol.*, **15**, 13–35.
WINDISH, W. W. & MAHTRE, N. S., (1965), 'Microbial amylases', *Adv. Appl. Microbiol.*, **7**, 273–298.
WINGARD, L. B., (1972), 'Enzyme engineering', *Adv. Biochem. Engin.*, **2**, 1–48.

6   Antibiotics

The observation that one microorganism could inhibit the growth of another had been made fairly frequently towards the end of the nineteenth century, and it had even been demonstrated that such an interaction might be mediated by the release from one organism of a metabolite which was toxic to the other. It was not, however, until the discovery and development of penicillin that a truly wide ranging search for antibiotics (as such compounds came to be known) was initiated.

The search has resulted in the discovery of an amazing number of antibiotic substances, the majority of which have proved in tests on laboratory animals to be too toxic for them to be of any practical clinical use.

Some of the antibiotic metabolites which have entered into clinical use are shown in Table 6.1.

Table 6.1   Some antifungal and antibacterial compounds produced by microorganisms

Active against	*Produced by*	
	Fungi	Bacteria
Fungi	Griseofulvin	Cycloheximide Amphotericin Pimaricin
Bacteria	Penicillins Cephalosporins Fusidic acid	Streptomycin Tetracycline Chloramphenicol Novobiocin Erythromycin Polymyxin Nisin

6.1   Penicillins

The penicillin story is something of a microbiological romance containing a number of chance episodes. The observation, in 1928, by Fleming of a stray fungal contaminant preventing the growth of staphylococci is a historical fact known to many. Perhaps not so widely appreciated is that Fleming, although noting that penicillin might be of therapeutic use, was initially interested in its usefulness in the isolation of *Haemophilus influenzae* from sputum in the presence of the fast growing Gram positive cocci which are always much more abundant than *Haemophilus* and tend to overgrow it on primary plate cultures. *Haemophilus* is relatively resistant to the compounds in Fleming's fungal brew whereas the Gram positive cocci are very sensitive.

The mould was called *Penicillium rubrum*, a misidentification which could have misled later workers had Fleming not taken care to preserve his isolate: it was later shown to be *Penicillium notatum*.

In 1932 Raistrick turned his attention to Fleming's mould confirming that an interesting antibiotic was liberated into the medium but he was unable to isolate the active substance. Raistrick and his colleagues were currently engaged in isolating, characterizing and cataloguing the wide diversity of mould metabolites and none at that time could appreciate that, not only was penicillin unstable, but that Fleming's isolate was only producing about $2\,\mu g/ml$ of an extremely active substance.

In 1938–39 Florey and Chain included *Penicillium notatum* in a study of naturally occurring antibacterial substances. Their work at Oxford was encouraging and by 1941 a quantity of material, so active that it was considered to be pure penicillin, had been isolated and was used for clinical trials. In fact, this preparation contained only about 2% penicillin; the rest was not only inactive but also, fortunately for the success of this story, non-toxic to mammals including man.

The Oxford group was working at a time when Britain was at war and, although the importance of the work was recognized, there appeared to be no money or materials to spare. The problem was taken to America and there given very high priority. The development of the industrial production of penicillin was essentially a three pronged study:

1. The search for other penicillin-producing strains in nature.
2. The laboratory improvement of strains by selection and mutation.
3. The optimizing of media and conditions of production.

The first approach led to the isolation from a melon of a strain of *Penicillium chrysogenum*, descendants from which are still used for the manufacture of penicillin. The second gave new mutants yielding considerably greater concentrations of penicillin (Table 6.2) and the third led to the realization that the nature and yield of penicillin could be influenced by the presence of precursors in the medium. Before penicillin had been purified and chemically characterized activity and yields were described in 'units', a concentration of 1 unit/ml being that which just inhibited growth of the Oxford strain of

Staphylococcus aureus. Once the complex nature of the penicillins was understood, that natural penicillin was a mixture of closely related compounds, and the individual components had been isolated, purified and chemically characterized, a unit could be redefined in terms of a mass of one of the pure crystalline penicillins:

$$1 \text{ unit} = 0.5988\,\mu g \text{ of sodium benzylpenicillin}$$

Not only was penicillin found to be a family of compounds but the exact nature of the mixture was shown to depend on the composition of the medium in which it was produced (Figure 6.1).

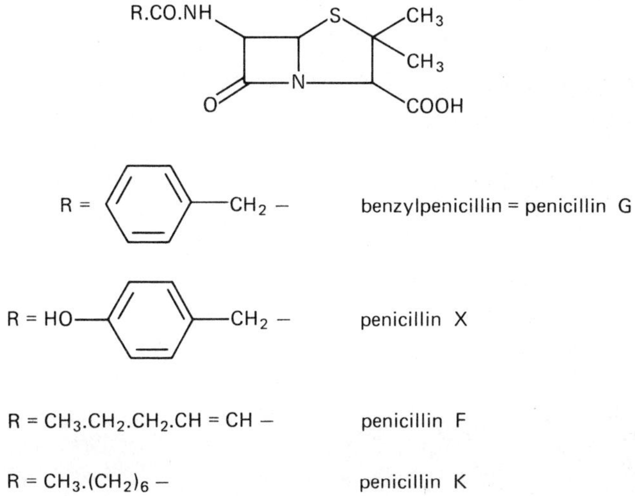

R = $\langle\!\!\langle\ \rangle\!\!\rangle$—CH$_2$ —   benzylpenicillin = penicillin G

R = HO—$\langle\!\!\langle\ \rangle\!\!\rangle$—CH$_2$ —   penicillin X

R = CH$_3$.CH$_2$.CH$_2$.CH = CH —   penicillin F

R = CH$_3$.(CH$_2$)$_6$ —   penicillin K

Figure 6.1   Naturally produced penicillins

The chemistry of the penicillins is complex and fascinating, but of immediate interest is the presence of two hydrolyzable nitrogen to carbonyl bonds (Figure 6.2). The imide bond of the four membered β-lactam ring is readily broken in both acid and alkaline media, and is also susceptible to hydrolysis by penicillinase, an enzyme synthesized by many bacteria.

Although the side chain amide bond is far more stable chemically, it can be hydrolyzed enzymically to yield 6-aminopenicillanic acid (6-APA) which has become an invaluable intermediate in the penicillin industry.

The natural penicillins, although highly active against sensitive bacteria, have three major limitations.

1. Their sensitivity to acid precludes oral administration since they are destroyed in the stomach.

Table 6.2 The progress in increasing yields of penicillin by a programme
of selection and mutation of strains of *Penicillium*

Date	Strain identification and origin	Yield (units/ml)	Comments
1929	*P. notatum* (Fleming)	2–20	Wild type isolate
1941	*P. notatum* (NRRL 832)	40–80	Wild type isolate
1943	*P. chrysogenum* (NRRL 1951)	80–100	Wild type isolate from a melon
	↓ *selection*		
1944	NRRL 1951 B25	100–200	(NRRL = Northern Regional Research Laboratories, Peoria)
	↓ *X-ray treatment*		
1944	X 1612	300–500	
	↓ *UV irradiation*		
1945	Q 176 (Wisconsin)	800–1000	
	↓ *UV irradiation*		
1947	BL 3D 10 (Wisconsin)	800–1000	Lacks the yellow pigment—chrysogenin
	↓ *nitrogen mustard*		
1949	49–133 (Wisconsin)	1500–2000	Poor growth
	↓ *nitrogen mustard*		
1951	51–20 (Wisconsin)	2400	Very poor growth
	↓ *selection*		
1953	53–399 (Wisconsin)	2700	Selected for improved growth
	↓ *selection*		
1960	Commercial strains	*c* 5000	Continuous selection for improved behaviour under commercial conditions
	? ⋮		
1970	Commercial strains	*c* 10,000	

2. Their sensitivity to penicillinase has reduced their usefulness in the
   treatment of infections.
3. Their spectrum of clinically useful activity is limited to Gram positive
   bacteria.

A very limited range of entirely synthetic side chain precursors can be
incorporated into the penicillin structure by *Penicillium chrysogenum*. Of
these, phenoxyacetic acid gives rise to penicillin V (Figure 6.3) which is
particularly stable to acid and was the first oral penicillin.

If starved of side chain precursors, *Penicillium chrysogenum* produces a very
low yield of 6-APA; on an industrial scale, it is far easier first to manufacture
benzylpenicillin and then split it to yield 6-APA. It is then possible to add,

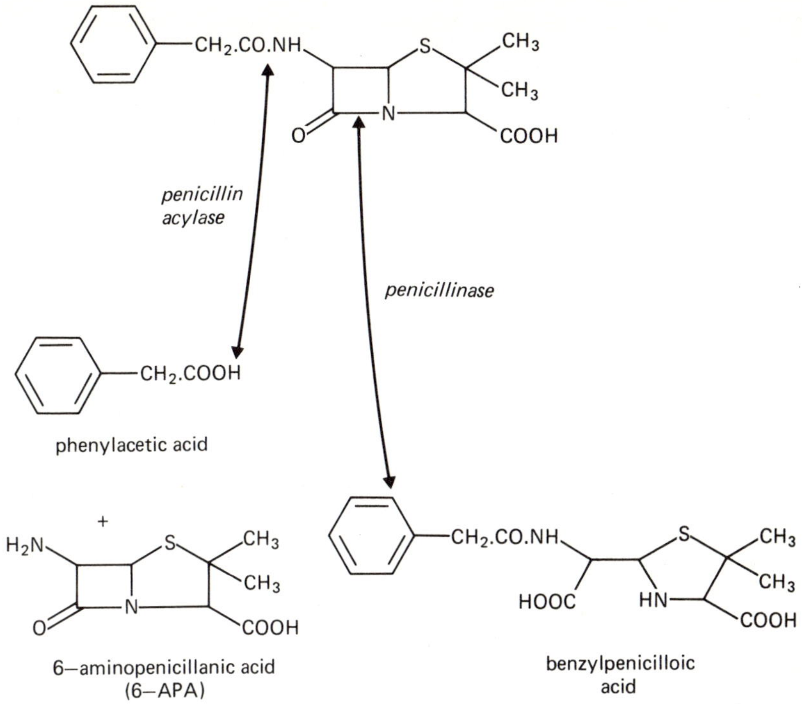

Figure 6.2   Enzyme breakdown of the penicillin molecule

chemically, a wide range of side chains to **6-APA**, producing synthetic penicillins which are either resistant to penicillinase, or active against Gram negative bacteria, or both.

Manufacture of benzylpenicillin requires a medium containing a carbon source, growth factors and a complete range of mineral salts. The carbon source first provides for biomass production and, later, for penicillin biosynthesis. Glucose is a useful carbon source for biomass production but is too readily assimilated, or wastefully metabolized, for the production of high yields of penicillin. At one time a mixture of glucose and lactose was incorporated into the medium, the former being readily assimilated to form biomass during the first stage of production, the latter being more slowly degraded and incorporated into the biosynthetic pathways leading to penicillin. However, lactose is an expensive sugar; its use has been replaced by the tightly controlled feeding of glucose to maintain a low concentration during the second phase of the production process.

The source of growth factors for industrial production is corn steep liquor, once a waste product, now a valuable byproduct of the maize starch industry.

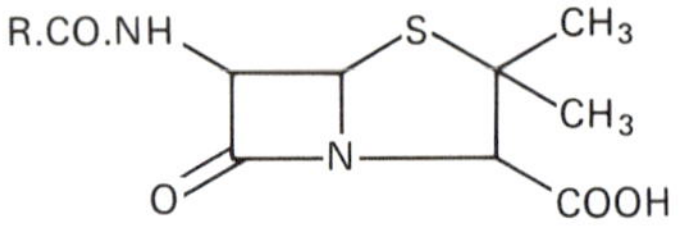

R	ACTIVITY
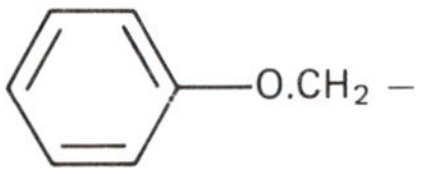 —O.CH$_2$— (phenoxymethyl)	Penicillin V, acid stable, penicillinase labile, active only against Gram +.
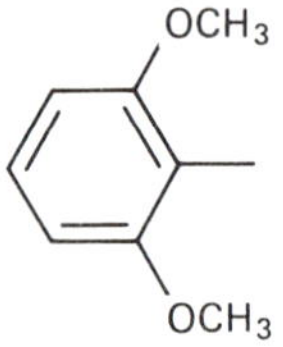 (2,6-dimethoxyphenyl) OCH$_3$ / OCH$_3$	Methicillin, very acid labile, penicillinase stable, active only against Gram +.
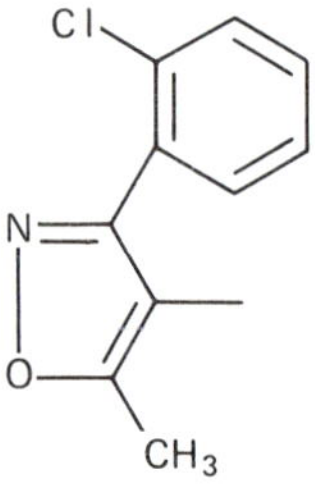 (3-(2-chlorophenyl)-5-methyl-isoxazol-4-yl)	Cloxacillin, acid stable, penicillinase stable, active only against Gram +.
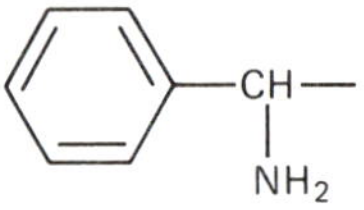 —CH— with NH$_2$ (α-aminobenzyl)	Ampicillin, acid stable, wide spectrum of activity against Gram + and Gram − with the notable exception of *Pseudomonas*.
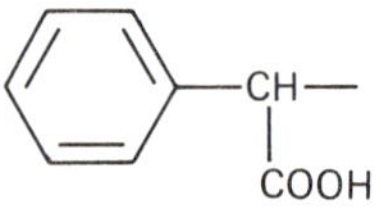 —CH— with COOH (α-carboxybenzyl)	Unstable but with a wide spectrum of activity including *Pseudomonas*. Carbenicillin.

Figure 6.3   Semisynthetic and synthetic penicillins

This material not only contains growth factors but also some useful precursors for penicillin biosynthesis. For example, certain breakdown products of phenylalanine are present and these are readily incorporated into the phenylacetyl side chain of benzylpenicillin. Indeed the early use of corn steep liquor by those working in America explains why American 'penicillin' produced during those pioneer days differed from material produced in Britain: the former contained a high level of penicillin G (benzylpenicillin) whereas the latter contained mainly penicillins F and K.

Corn steep liquor does not, however, provide sufficient levels of such precursors to satisfy the biosynthetic potential of a modern strain of *Penicillium chrysogenum* and phenylacetic acid, or a derivative, is thus fed in a controlled manner during the later stages of production. Careful control of the addition of phenylacetic acid is essential so that its concentration does not build up to a level which would be toxic to the fungus.

Once production is complete, the mycelium is removed on a rotary vacuum filter drum and the harvested liquors cooled, acidified and rapidly extracted with a solvent such as methyl isobutyl ketone. The penicillin may be carefully back extracted with aqueous sodium bicarbonate and the resulting sodium salt of benzylpenicillin recovered by spray drying.

To obtain the valuable intermediate, 6-APA, a solution containing up to 15% benzylpenicillin is treated with a preparation of penicillin acylase. This enzyme can be obtained, in a cell-bound form, by growing strains of *Escherichia coli* aerobically on a protein-rich medium and inducing enzyme formation with phenylacetic acid. Deacylation occurs optimally at pH 8·5 and this condition is maintained by the continuous addition of alkali to neutralize the phenylacetic acid produced during the reaction. Monitoring of the quantity of alkali added also provides information on how the reaction is proceeding.

After deacylation the cells and their bound enzyme are removed by centrifugation, giving a concentrated solution of the sodium salts of phenylacetic acid and 6-APA. The phenylacetic acid is recovered by solvent extraction after acidifying and 6-APA, an amphoteric compound, can be precipitated by carefully adjusting the pH to its isoelectric point, 4·2, and cooling.

6.2   Cephalosporins

Some strains of mould of the genus *Cephalosporium* produce a mixture of antibiotics loosely referred to as the cephalosporins. In fact, the compound called cephalosporin P is a steroid-like antibiotic and that termed cephalosporin N is a derivative of 6-APA and hence is a penicillin; it is cephalosporin C (Figure 6.4) which is the parent compound of the family of antibiotics now referred to as the cephalosporins. By careful hydrolysis the α-aminoadipyl side chain can be removed to yield 7-aminocephalosporanic acid (7-ACA); this can then be converted to a range of derivatives with different side chains in a manner analogous to the production of the synthetic penicillins.

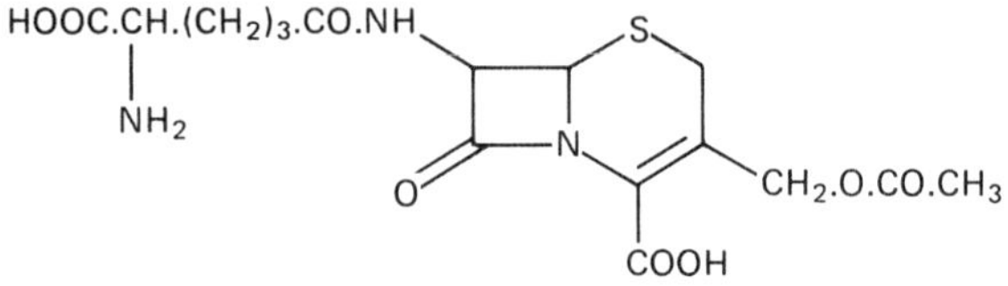

$$\text{HOOC.CH.(CH}_2)_3\text{.CO.NH} \quad\quad S$$

Figure 6.4   Cephalosporin C

Although closely related to penicillins, the derivatives of 7-aminocephalosporanic acid are resistant to penicillinases (though susceptible to hydrolysis by another group of enzymes, the cephalosporinases). The penicillins are remarkable for their lack of toxicity but there is an increasing occurrence of allergic responses to them. A person sensitized to one penicillin frequently becomes allergic to others but not to the cephalosporins.

The natural antibiotic, cephalosporin C, is not widely used clinically; it is, however, employed in the manufacture of 7-ACA which is then reacylated with a number of alternative side chain acids to yield the commercially available, synthetic cephalosporins.

6.3   Streptomycin

The concept established by the discovery of penicillin, that one microorganism might produce relatively small molecular weight metabolites active in the destruction or inhibition of other microorganisms, naturally led to deliberate searches for other organisms and metabolites. Waksman and his team, arguing from the ecological consideration that one of the factors involved in the fierce competition of microorganisms living in the soil might be antibiotic production, started in 1939 to screen soil microorganisms. They discovered many organisms producing compounds with antibiotic activity; none of these compounds proved to have any medical potential until, five years and some 10,000 isolates later, streptomycin was isolated from a strain of *Streptomyces griseus* cultured from heavily manured soil. This species is common in soil, but streptomycin production appears to be highly strain-specific and many industrial processes still use strains directly derived from Waksman's original isolate.

The molecular structure of streptomycin (Figure 1.17) contains three units, each synthesized from a molecule of glucose. Industrial media for streptomycin production contain materials such as extracted soybean meal as a carbon and nitrogen source for biomass production, distillers' solubles as a source of growth factors and glucose as a major precursor of streptomycin itself. One interesting role of distillers' solubles is the early induction of an α-mannosidase because of the presence of yeast mannans. As a precursor during biosynthesis of streptomycin an α-mannoside (Figure 1.18) of low biological activity is formed which is subsequently hydrolyzed by α-mannosidase to the required antibiotic. Streptomycin is produced at the end of the exponential phase of growth and during the stationary phase and, as the stationary phase progresses, the filamentous form of the organism breaks down making filtration and subsequent recovery of the antibiotic more difficult. It is thus desirable to produce a good yield of streptomycin as early as possible and this will depend, in some measure, on the early induction of the α-mannosidase which is subject to catabolite repression while glucose remains in the medium (see Chapter 1).

Streptomyces griseus is an obligate aerobe and the process requires efficient aeration. One of the major difficulties in the large scale manufacture of streptomycin is possible contamination by actinophage causing premature lysis of the streptomyces cells. If contamination occurs during a seed stage the culture is rendered useless as an inoculum for the next stage. If it occurs during the production stage, streptomycin yield is reduced and, because of considerable lysis, recovery of the product is difficult. Once production is complete, the branching filaments of the streptomyces are removed by filtration, usually with the help of a filter-aid because the cells are just beginning to break down. The antibiotic is then absorbed from the filtrate onto charcoal columns. Some of the impurities, which are also absorbed, can be eluted with organic solvents. The streptomycin itself is eluted from the column with an acid, making use of the fact that the compound is a strong base readily forming water soluble salts. Alternatively, cationic resin columns may be used.

Streptomycin quickly emerged as a valuable antibiotic for the treatment of infections caused by certain Gram-negative bacteria which were resistant to penicillin. Today it is one of the principal drugs used in the treatment of tuberculosis: it has to be alternated with other compounds such as isonicotinic acid hydrazide (isoniazide) because of the ease with which *Mycobacterium tuberculosis* acquires streptomycin resistance, and because prolonged streptomycin administration can give rise to serious vertigo and deafness.

6.4   The tetracyclines

In the late 1940s the principal antibiotics available were the natural penicillins and streptomycin. Although the latter is stable to the acidic conditions in the stomach, and thus found a use in the treatment of Gram-negative enteric infections, it is not absorbed when administered orally; both compounds had thus to be given by injection. The search for new antibiotics was intense and members of the genus *Streptomyces* were particularly well studied. In 1948, two wide spectrum oral antibiotics, aureomycin and chloramphenicol, were discovered.

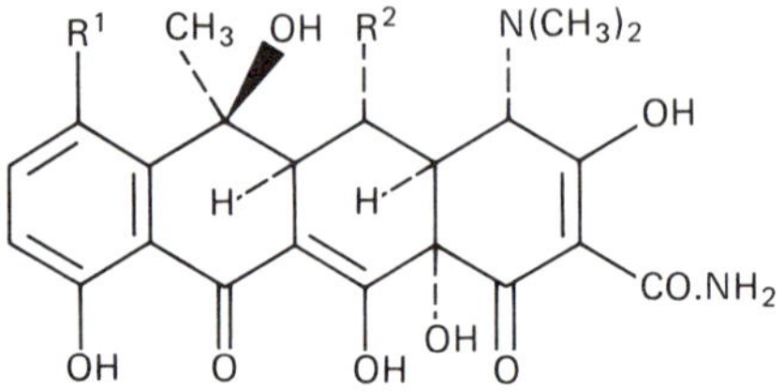

Figure 6.5 The tetracyclines

R^1	R^2	
H	H	tetracycline
C1	H	chlortetracycline (aureomycin)
H	OH	oxytetracycline (terramycin)

Aureomycin (chlortetracycline: Figure 6.5) was the first member of the group of antibiotics now referred to as the tetracyclines. The subsequent tetracyclines (tetracycline itself and oxytetracycline) have almost identical ranges of antibiotic activity, but their production was subject to different patents.

A typical medium for the production of tetracyclines, such as chlortetracycline, may contain starch, ground nut meal, corn steep liquor, ammonium sulphate, beet molasses, calcium carbonate and, in the case of chlortetracycline itself, an additional source of chloride ions such as potassium chloride.

Although calcium carbonate plays a part in maintaining a suitable pH for growth of the producing organism, *Streptomyces aureofaciens* for chlortetracycline and *St. rimosus* for oxytetracycline, it has a further important role. Concentrations of greater than 100 mg tetracycline/l have a profound effect on the growth of the producing organism. The presence of the poly-enol-ketone structure in the molecule of the tetracyclines makes them powerful chelating agents and the consequent removal of essential trace elements is probably a major factor in the inhibition of growth. The excess calcium in the medium continuously removes the tetracycline as an insoluble calcium complex and yields in a modern process may be in excess of 3 g/l.

The anitbiotic may be recovered by completing the precipitation with lime, filtering, and extracting the filter cake with dilute sulphuric acid. The soluble tetracycline sulphate is readily separated from the insoluble calcium sulphate. One method of purifying the tetracycline further is to saturate the solution of the sulphate with butanol and sodium chloride. On cooling to 2–5°C, tetracycline hydrochloride of high purity crystallizes out.

The first two members of the family isolated, chlortetracycline and oxytetracycline, have a number of disadvantages for medical use. They are not readily absorbed from the alimentary tract and a number of gastro-intestinal upsets may occur following prolonged oral administration. Tetracycline itself does not have these side effects to the same extent and the chemically derived compounds, methacycline and doxycycline (Figure 6.6), are even more readily absorbed allowing the use of lower doses.

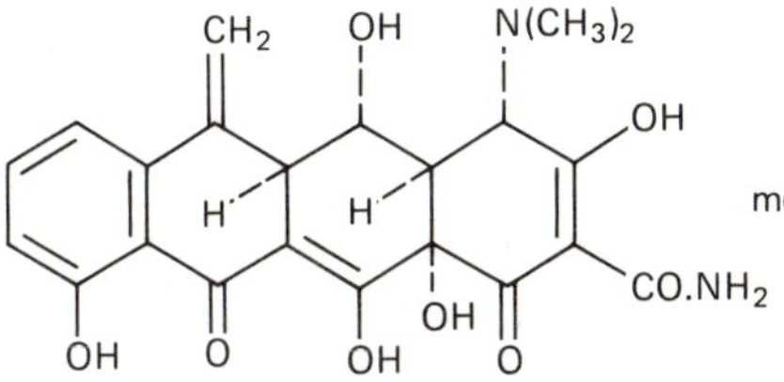

Figure 6.6   Methacycline and doxycycline

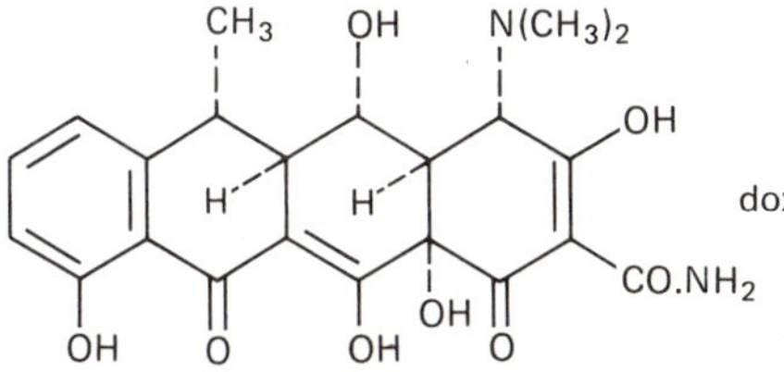

Tetracycline itself may be produced by some strains of *St. aureofaciens*, especially if chloride ions are completely removed from the medium, but yields are not high and it is expensive to remove chloride completely from the complex media used commercially. Tetracycline can be manufactured chemically from chlortetracycline by catalytic hydrogenation using a palladium-charcoal catalyst and an organic base, such as triethylamine, to neutralize the hydrochloric acid formed. Alternatively, laboratory derived mutants of *St. aureofaciens* and wild-type isolates of *St. lusitanus* which produce a good yield of tetracycline, even in the presence of chloride ions, have been obtained.

Although the biosynthetic pathway to the tetracyclines is complex in detail, and not yet fully worked out, the tetracyclic carbon skeleton is derived from acetate units by the polyketide route (Figure 6.7) with the addition of one C-methyl and two N-methyl groups via methionine from the C_1 pool.

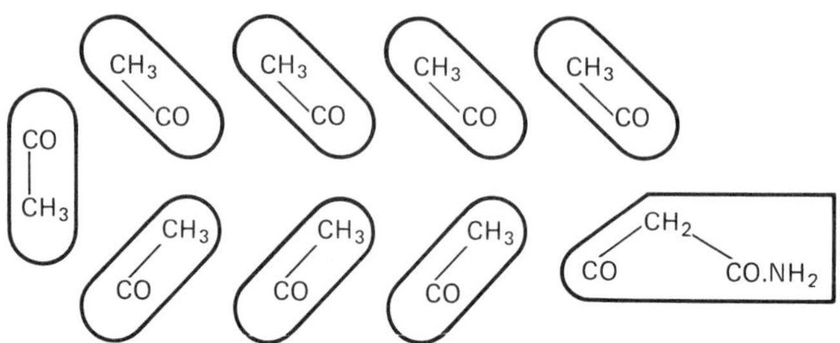

Figure 6.7   The biosynthesis of the tetracycline skeleton from a malonamoyl unit and eight acetate units

6.5   Chloramphenicol (= chloromycetin)

This compound was isolated from strains of *Streptomyces venezuelae* at the same time that the tetracyclines were first obtained. It is a broad spectrum antibiotic having activity against rickettsias as well as some groups of viruses. At the time following its discovery it was the potent activity against such Gram-negative pathogens as *Salmonella typhi*, *Haemophilus influenzae* and *Bordetella pertussis* which particularly excited interest.

Chloramphenicol is unusual amongst antibiotics in having a chemical structure (Figure 6.8) simple enough to make its manufacture by chemical processes very competitive with its production by fermentation. Because of its toxicity to the producing organism yields by direct fermentation are restricted to less than 100 mg/l irrespective of the medium used. It is a very stable, neutral compound readily extracted from fermentation liquors by industrially available organic solvents, from which it can be recovered by concentration and crystallization. Because of its stability, and almost complete absorption from the alimentary canal, chloramphenicol is an ideal oral antibiotic and had been widely used for a number of years before it was recognized as being responsible for two toxic effects in man. The 'grey syndrome' in babies and the irreversible reduction of bone marrow activity in adults are both potentially lethal results of prolonged administration of chloramphenicol. Despite these unfortunate effects the antibiotic is still used in many parts of the world though, hopefully, with an increasing awareness and understanding of its toxicity.

6.6   Novobiocin

The considerable biosynthetic versatility of the streptomycetes is demonstrated by comparing the chemical structure of the antibiotics which they can produce (see Figure 6.8). Novobiocin is a particularly complex metabolite for

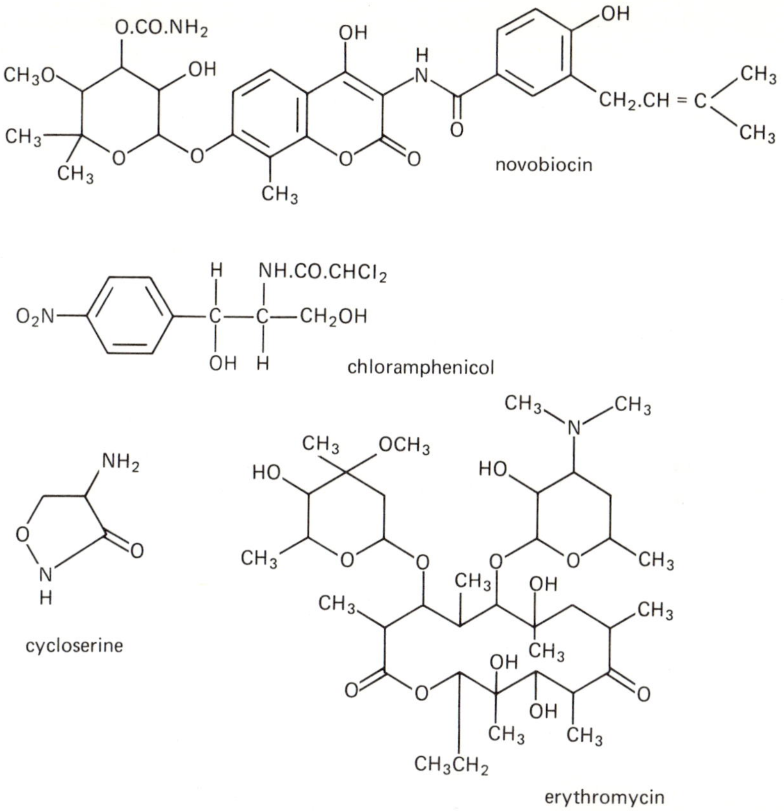

Figure 6.8   Structures of selected antibiotics produced by streptomycetes

it would seem to be composed of three parts each of different biosynthetic origin.

The search for new antibiotics has been so intense in the past that several of them were discovered at approximately the same time by a number of groups working independently. Thus novobiocin was isolated from *Streptomyces spheroides* three times during the two years 1955 and 1956 and was consequently known by several different names—cathomycin, streptonivicin and cardelmycin—before its present name was internationally accepted. It aroused interest because of its high activity against *Staphylococcus aureus* and, because it does not show cross-resistance with other groups of antibiotics, is particularly useful against those strains which have acquired resistance to penicillins and tetracyclines. However, many strains of *Staphylococcus aureus* readily acquire resistance to novobiocin too.

Novobiocin can be manufactured in deep fermentation using a medium containing distillers' solubles and glucose with good aeration and stirring. It is necessary to maintain the pH at about 8·5 for, like many antibiotics, novobiocin is toxic to the producing organism but this toxicity is pH-dependent and increases with decreasing pH.

When production is complete the fermentation liquors are adjusted to pH 8·0 at which novobiocin has a high solubility in the aqueous phase. Mycelium is removed by filtration with a filter aid, the filtrate adjusted to pH 6 and extracted with an organic solvent such as amyl acetate. Novobiocin is then back-extracted with an aqueous buffer at pH 10 from which it can be either precipitated, or re-extracted into amyl acetate, following acidification.

6.7   Erythromycin

This antibiotic, with oleandomycin and spiramycin, belongs to a family of compounds referred to as macrolides on account of the presence of a macrocyclic lactone ring in the structure. Erythromycin itself is produced by *Streptomyces erythreus* on a medium rich in carbohydrates. Production of the antibiotic occurs after the nitrogen source used for biomass production has become depleted, excess carbohydrate being then converted into erythromycin during the stationary phase. This common pattern of antibiotic biosynthesis may require the use of at least two different media for optimum production, and possibly a third if the initial inoculum is to be a spore suspension. To illustrate this the composition of three media that can be used

Table 6.3   The composition of media which can be used in the production of erythromycin (quantities are expressed in g/l)

Sporulation medium	Medium for the production of vegetative inoculum	Medium for erythromycin production
Starch 5·0	Starch 15·0	Glucose 40·0
Glucose 5·0	Soybean meal 15·0	Corn extract 10·0
Tryptone 5·0	Corn steep liquor 5·0	$(NH_4)_2SO_4$ 6·0
Betaine 0·5	NaCl 5·0	NaCl 2·5
Curbay B.G. 2·0	$CaCO_3$ 3·0	$CaCO_3$ 10·0
K_2HPO_4 0·2		Sperm oil 6·0
NaCl 10·0		
$CaCl_2$ 0·08		
Mineral mixture 2·0 ml		
Agar 20·0		

in the production of erythromycin is shown in Table 6.3. The first allows the production of a good spore suspension which is used to inoculate the second in which vigorous vegetative growth is possible. This second stage thus provides a vegetative inoculum for the third medium in which optimum erythromycin production occurs.

The antibiotic can be extracted from the culture filtrate with amyl acetate after adjusting the pH to 9·0. It can then be back extracted into aqueous buffer at pH 4·0 and this aqueous extract concentrated under reduced pressure, adjusted to pH 9·0, and cooled when the free base crystallizes out.

The biosynthesis of erythromycin involves the building up of the macrolide skeleton from propionate subunits (Figure 6.9), themselves produced, via methylmalonyl coenzyme A, from oxaloacetate which is a readily available intermediate from carbohydrate metabolism.

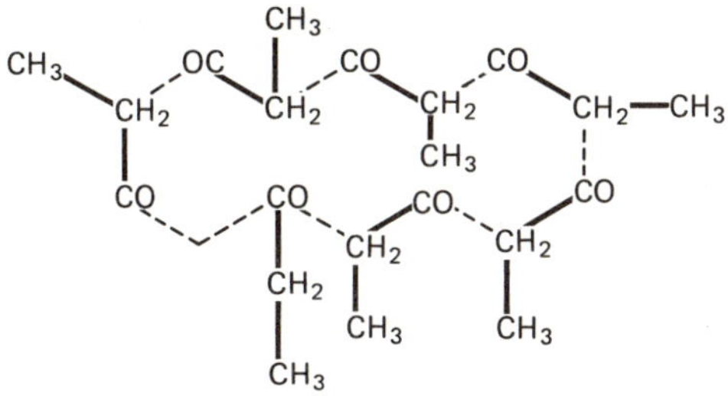

Figure 6.9   The biosynthesis of the macrolide skeleton of erythromycin from propionate subunits

6.8   Polypeptide antibiotics produced by species of *Bacillus*

A strain of *Bacillus brevis*, isolated from soil by Hotchkiss and Dubos, produces a mixture of antibacterial compounds: an active precipitate is obtained by acidifying an autolyzed culture. This precipitate is readily fractionable into (1) inactive proteins, which are soluble in neutral buffer and (2) alcohol-soluble, antibacterial polypeptides, referred to collectively as tyrothricin. Tyrothricin can be fractionated further into two major groups of antibiotics, the tyrocidins and the gramicidins, both of which contain some D-amino acid residues as well as the normal L-isomers.

This group of compounds is generally active against Gram positive bacteria only and is restricted in use to incorporation into lozenges for the treatment of mouth and throat infections.

Of greater therapeutic importance is a group of strongly basic polypeptide antibiotics, referred to as the polymyxins, produced by *Bacillus polymyxa*.

Polymyxins and bacitracins

These compounds are widely active against Gram negative bacteria, although strains of *Proteus* are resistant; they are of great clinical importance because of their activity against *Pseudomonas aeruginosa*, a bacterium common in hospital infections, particularly in burns units and urinogenital wards, and highly resistant to most other antibiotics. They are unfortunately strongly nephrotoxic and, of the five readily available compounds, only polymyxins B and E have a low enough risk of kidney damage for clinical use. The structure of polymyxin D is shown in Figure 6.10 from which it can be seen that the strongly basic nature of these antibiotics is due to the large number of residues of diaminobutyric acid in the molecule.

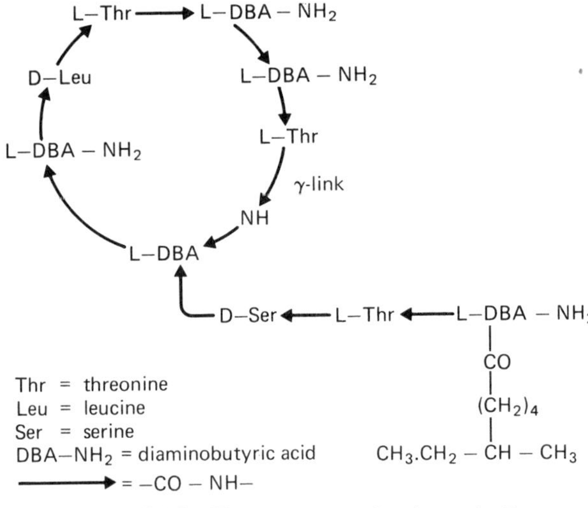

Figure 6.10   The structure of polymyxin D

Colistin, subsequently shown to be identical to polymyxin E, was isolated from another strain of *Bacillus* by Japanese workers.

The polymyxins are produced in a highly aerobic process using a medium containing glucose, yeast extract and inorganic salts. The bacteria are removed by filtration, employing a filter aid, and the polymyxins absorbed from the filtrate using activated charcoal. Being strongly basic they are readily eluted with methanolic hydrochloric acid. The hydrochloride may then be precipitated with acetone and dried under vacuum to give a crude preparation of the antibiotic. The crude hydrochloride may be further purified by extracting with butanol at pH 9·0 and back extracting with acid, followed by chromatography and crystallization of a highly crystalline derivative such as the naphthalene β-sulphonic acid salt.

Another cyclic polypeptide is bacitracin which, as a zinc complex, is useful for topical application in the treatment of superficial pyogenic infections. Bacitracin (Figure 6.11) is produced by strains of *Bacillus subtilis* and *B. licheniformis* in an aerobic process at 37°C using a medium typically containing soybean flour, starch, calcium carbonate and magnesium sulphate. Bacitracin can be precipitated from the culture filtrate by a number of agents including the dye Polar Yellow 5G, ammonium molybdate (at pH 2·0), alkali soluble lignins and lignosulphonates (at pH 3·0) and zinc ions. The zinc complex has better stability in storage than does bacitracin itself and crude preparations have been used as additives to animal feeds. For medical use the precipitated form of bacitracin is resuspended in water and extracted into a

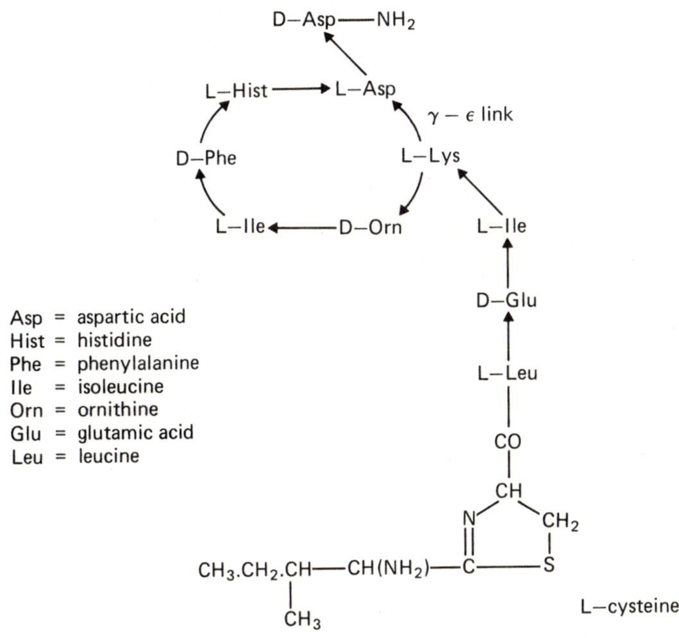

Figure 6.11   The structure of bacitracin A, the principal active component of the commercial product

solvent such as butanol after adjusting to a suitable pH. It is then repre-cipitated after back-extraction into an aqueous buffer.

6.9   Antifungal antibiotics

From the mycelial fungi themselves, the only clinically useful antifungal metabolite is griseofulvin (Figure 6.12) produced by *Penicillium griseofulvum* (= *P. urticae*).

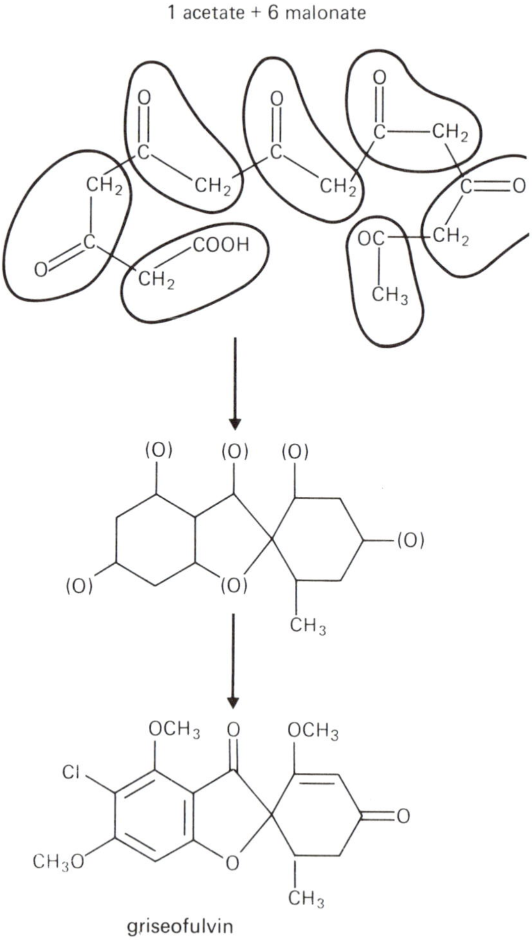

Figure 6.12   Structure and biosynthesis of griseofulvin

Griseofulvin is essentially fungistatic rather than fungicidal but, because of its remarkably low toxicity to mammals and the fact that after oral administration it accumulates in the skin, nails and hair, it is useful in the treatment of the superficial dermatomycoses, such as ringworm and favus.

Yields of up to 1·5 g/l may be obtained after seven days at 25°C using a medium containing corn steep liquor, lactose, calcium carbonate and potassium dihydrogen phosphate. The chlorine atom in the structure of griseofulvin is provided by potassium chloride added to the medium. Lactose, though not readily assimilated by the mould, acts as a continuous low level supply of a carbon source for griseofulvin biosynthesis.

Griseofulvin is a perfect example of a secondary metabolite, being produced during the second stage of fungal development, and is a model structure for demonstrating the polyacetate, or polyketide, route of biosynthesis (Figure 6.12). Every oxygen function can be seen to be derived from the carboxyl group of an acetate precursor.

The isolation of griseofulvin from the culture medium is relatively straightforward for it is readily extracted into solvents, purified and crystallized.

Several quite different antifungal compounds can be isolated from the streptomycetes. Strains of *Streptomyces griseus* are used for the manufacture of cycloheximide (=actidione). Although not widely used clinically this compound is frequently included in the formulation of selective media for the isolation and enumeration of bacteria, in order to inhibit the growth of fungi. Cycloheximide (Figure 6.13) is readily isolated by solvent extraction and can be obtained in a high degree of purity by chromatography on columns of activated charcoal.

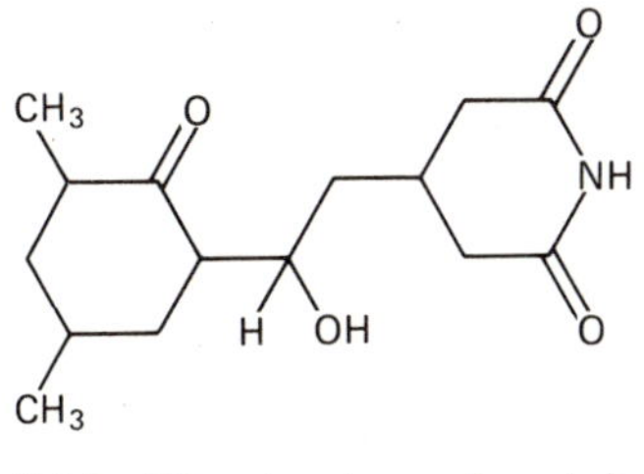

Figure 6.13   The structure of cycloheximide

A far more complex group of antifungal compounds produced by streptomycetes comprises the polyene macrolides. One of the chemically simplest of this group is pentamycin (filipin; Figure 6.14). Pentamycin is a neutral molecule, but the majority of polyene macrolides, including the clinically useful amphotericin, are complex amphoteric compounds containing an amino sugar attached to one of the many hydroxyl groups.

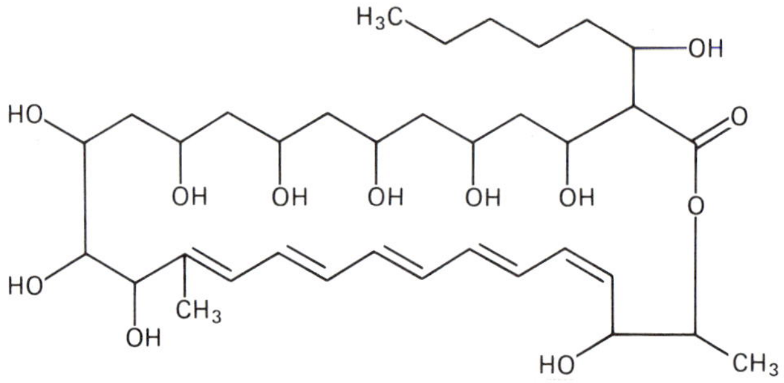

Figure 6.14   The structure of pentamycin

The number of polyene macrolides known is now so large as to require classification: there are at least two approaches. The first makes use of the presence in these compounds of a number of conjugated double bonds ranging from four to seven. This feature gives rise to an easily recognizable pattern of absorptions in their ultra violet spectra. In particular there is a pattern of three distinct peaks at the long wavelength end of the spectrum the position of which depends on the number of conjugated double bonds (Table 6.4).

Table 6.4   The position of three characteristic peaks in the UV spectra of polyene macrolides

| | *Position of peak* | | |
| *Group* | (wavelength in nm) | | |
	I	II	III
Tetraene	290–295	303–309	318–323
Pentaene	318–324	333–338	348–358
Hexaene	339–341	356–358	377–380
Heptaene	358–372	376–392	399–415

Another way of classifying the polyene macrolides is on the basis of their behaviour as acids, bases, neutral or amphoteric compounds (Table 6.5).

A typical medium for polyene macrolide production may contain glucose as the carbon source, soybean meal as the nitrogen source, corn steep liquor as a source of growth factors, and additional inorganic salts. The producing strains are all obligate aerobes and cultures are usually maintained at 26–30°C.

The polyene macrolides are not normally excreted into the medium but are contained within the biomass: the streptomycete mycelium must be harvested

Table 6.5 Polyene macrolides which are, or have been, manufactured commercially

Compound	Number of conjugated double bonds	Ionic status	Producing organism (Streptomyces)	LD_{50} (mg/kg) (mice)
Amphotericin B	7	Amphoteric	S. nodosus	280
Candicidin	7	Amphoteric	S. griseus	79
Nystatin*	4	Amphoteric	S. aureus	20–26
Pentamycin**	5	Neutral	S. pentaticus	16–33
Pimaricin	4	Amphoteric	S. natalensis	650
Trichomycin***	7	Amphoteric	S. hachijoensis	2

* = Fungicidin
** = Fumichromin = Lagosin = Filipin
*** = Hamycin

and extracted, a suitable solvent being *n*-butanol saturated with water. Because these compounds have a low solubility in most readily available solvents, are unstable to acids, alkalis and light and are difficult to crystallize, their purification is a tedious process.

The polyene macrolides are particularly active against yeast-like fungi such as *Candida albicans* although amphotericin B is also active against some filamentous fungi, such as *Aspergillus fumigatus*, which may also be involved in mycotic infections.

6.10   Non-medical uses of antibiotics

As the technology of antibiotic manufacture improved, and many changed from scarce and expensive commodities to readily available and relatively cheap compounds, so the possibility of non-medical use has arisen. Two such uses which have been explored are as food preservatives and as animal feed additives. The latter use arises from the observation that the addition of low concentrations of certain antibiotics to animal feeds often results in increased weight gain, especially under conditions of intensive rearing. Concern about the long term effects of such use, both from the point of view of chronic toxicity to man, and the increasing occurrence of infective multiple drug resistance factors in bacteria, has led to increasing control of these non-medical uses of antibiotics. Thus the use of antibiotics as food preservatives is now forbidden in Great Britain except in three instances: up to 5 ppm of chlortetracycline or oxytetracycline may be added to the ice used for the storage of raw fish; nisin may be added with no limit to cheeses (in which it it often occurs naturally being a product of *Streptococcus lactis*) and to canned foods; nystatin may be used on the skin, but not in the flesh, of bananas.

Bibliography

ABRAHAM, E. P. & NEWTON, G. G. F., (1965), 'The cephalosporins', *Adv. in Chemotherapy*, **2**, 23–85.

BERDY, J., (1974), 'Recent developments of antibiotic research and classification of antibiotics according to chemical structure', *Adv. Appl. Microbiol.*, **18**, 309–406.

COLE, M., (1966), 'Microbial synthesis of penicillins', *Process Biochemistry*, **1**, 334–338, 373–377.

GOLDBERG, H. S., (1964), 'Non-medical uses of antibiotics', *Adv. Appl. Microbiol.*, **6**, 91–117.

HAMILTON-MILLER, J. M. T., (1974), 'Fungal sterols and the mode of action of the polyene antibiotics', *Adv. Appl. Microbiol.*, **17**, 109–131.

HESSELTINE, C. W. & ELLIS, J. J., (1975), 'Antibiotic-producing fungi: current status of nomenclature', *Adv. Appl. Microbiol.*, **19**, 47–57.

HICKEY, R. J., (1964), 'Bacitracin, its manufacture and uses', *Progress in Industrial Microbiol.*, **5**, 93–150.

HOEKNEMA, H. & SMITH, C. G., (1961), 'Novobiocin', *Progress in Industrial Microbiol.*, **3**, 91–139.

INGRAM, J. M. & BLACKWOOD, A. C., (1970), 'Microbial production of phenazines', *Adv. Appl. Microbiol.*, **13**, 267–280.

JUKES, T. H., (1973), 'Public health significance of feeding low levels of antibiotics to animals', *Adv. Appl. Microbiol.*, **16**, 1–28.

KAVANAGH, F. (Ed.), (1972), *Analytical Microbiology*, Vol. 2, Academic Press.

LECHEVALIER, H. A., (1975), 'Production of the same antibiotics by members of different genera of microorganisms', *Adv. Appl. Microbiol.*, **19**, 25–45.

LESSEC, E. F., (1975), 'Current status of nomenclature of antibiotic-producing bacteria', *Adv. Appl. Microbiol.*, **19**, 71–76.

MALIK, V. S., (1972), 'Chloramphenicol', *Adv. Appl. Microbiol.*, **15**, 297–331.

MOSS, M. O., (1977), 'Enzymic alterations of penicillins and cephalosporins', in *Topics in Enzyme and Fermentation Biotechnology 1*, Ed. A. Wiseman, Ellis Horwood.

OWENS, D. R., LUSCOMBE, D. K., RUSSELL, A. D. & NICHOLLS, P. J., (1975), 'The cephalosporin group of antibiotics', *Adv. in Pharmacol. Chemother.*, **13**, 83–170

PERLMAN, D., (1967), 'Production of polyene antifungal agents by streptomycetes', *Progress in Industrial Microbiol.*, **6**, 1–19.

PORTER, J. N., (1971), 'Prevalence and distribution of antibiotic producing actinomycetes', *Adv. Appl. Microbiol.*, **14**, 73–90.

PRICE, K. E., GODFREY, J. C. & KAWAGUCHI, H., (1974), 'Effect of structural modifications on the biological properties of amino-glycoside antibiotics containing 2-deoxy streptamine', *Adv. Appl. Microbiol.*, **18**, 191–307.

RICHMOND, M. H. & SYKES, R. B., (1973), 'The β-lactamases of gram negative bacteria and their possible physiological role', *Adv. in Microbial Physiol.*, **9**, 31–85.

ROLINSON, G. N. & SUTHERLAND, R., (1973), 'Semisynthetic penicillins', *Adv. in Pharmacol. Chemother.*, **11**, 152–212.

SADOFF, H. L., (1972), 'The antibiotics of *Bacillus* species: their possible roles in sporulation', *Progress in Industrial Microbiol.*, **11**, 1–27.

SENSI, P. & THIEMANN, J. E., (1967), 'Production of rifamycin', *Progress in Industrial Microbiol.*, **6**, 21–60.

STARK, W. M. & SMITH, R. L., (1961), 'The erythromycin fermentation', *Progress in Industrial Microbiol.*, **3**, 211–230.

VANDAMME, E. J. & VOETS, J. P., (1974), 'Microbial penicillin acylases', *Adv. Appl. Microbiol.*, **17**, 311–361.

7 Alcoholic Beverages

The art of making alcoholic beverages by fermentation must have been discovered many times in man's history, for such beverages occur in many different forms the world over. The raw material varies according to what is regionally available, but usually comprises either sugary materials (fruit juices, plant saps, honey) or starches (grain or root) which have to be saccharified by hydrolysis using such amylase sources as malted grain or amylolytic micro-organisms. Table 7.1 summarizes the substrates for some major and minor alcoholic drinks of the world.

The result of allowing fermentation of such substrates is a liquid containing anything from a few percent up to 16 % or more of ethanol, acid in pH (either because of the acidity of the substrate or because of some concomitant lactic acid fermentation) and depleted of nutrients for most microorganisms; all these factors combine to confer a degree of biological stability on the product. The product may be drunk fresh or, more usually, after a period of ageing, or it may be distilled to increase alcoholic strength and modify flavour and drunk as a spirit.

7.1 Alcoholic fermentation

The fermenting organism is usually the yeast *Saccharomyces cerevisiae*, or some variant or closely related form. Ale yeasts are *S. cerevisiae* (top fermenting), lager yeasts are a bottom-fermenting variant termed *S. carlsbergensis*, wine yeasts are *S. cerevisiae* var. *ellipsoideus* (often shortened to *S. ellipsoideus*, cider yeast is *S. uvarum*, saké yeast has been termed *S. saki* and that of the pulque fermentation, *S. carbajali*, and so on. Traditional and domestic-scale fermentations may utilize mixed yeast fermentations including, for example, *S. intermedius*, and even bacterial fermentations leading to alcohol production as in the Indonesian tuwak (*Zymomonas* sp.).

Among the alcohol-producing yeasts, ability to hydrolyze oligo- and di-saccharides varies, though most can split sucrose and maltose. All can assimilate and utilize hexose monosaccharides such as glucose and fructose, metabolizing them by the Embden-Meyerhof-Parnas pathway to produce pyruvic acid (see Figure 7.1); a final stage produces ethanol.

Table 7.1 Substrates for selected alcoholic beverages (non-distilled)

Substrates	Beverage	Country	Saccharifying agent
Starchy			
(Barley + other cereals)	Ale	Belgium W. Germany Canada Australia	Barley malt
	Lager	Worldwide (industrial countries)	Barley malt
Barley, rye, rice, beet	Kvass	USSR	Barley and rye malt
Millet	Busa	USSR (Crimea)	
	Braga	Roumania	
	Thumba	India	
Rice	Arak*	India, S.E. Asia	
	Busa	Turkestan, SSR	
	Pachwai	India	*Mucor* sp.
	Saké	Japan	*Asp. oryzae*
	Sonti	India	*Rhizopus* sp.
Rice (red)**	Anchu	Taiwan	
	Hung-Chu	China	
Sorghum	Kaffir beer	Malawi	Sorghum malt *Aspergillus* sp. *Mucor rouxii*
	Merissa	Sudan	*Bacillus* spp.
Sweet potato	Awamori	Japan	
Sugary			
Agave spp. (sap)	Pulque	Mexico	
Apple (juice)	Cider	UK, France, N. America	
Grape (juice)	Wine	Temperate; N. and S. Hemispheres	(not required)
Honey	Mead	UK	
Pear (juice)	Perry	UK, France	
Palmyra (juice)	'Toddy'***	India, S.E. Asia	
Palm flower-stalk (juice)	Tuwak	Indonesia	

* With molasses, palm juice; chiefly distilled.
** Rice fermented with the pigmented *Monascus purpureus*.
*** Or distilled as arak.

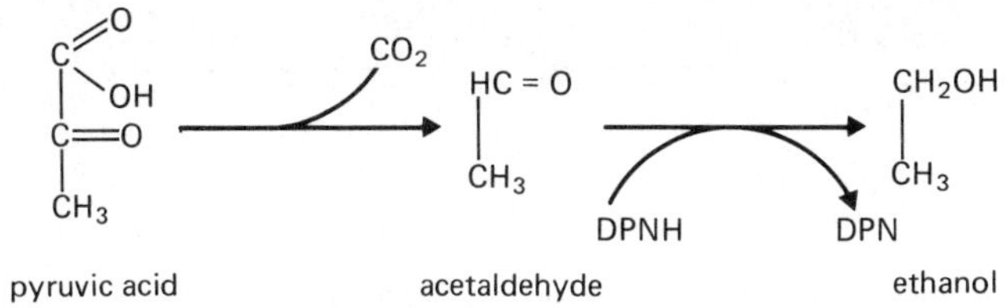

Figure 7.1

The overall reaction can be summarized:

$$C_6H_{12}O_6 + 2P + 2ADP \rightarrow 2C_2H_5OH + 2CO_2 + 2H_2O + 2ATP$$

The net energy gain for the yeast derives from the conversion of two ADP molecules to two ATP molecules. The free energy of hydrolysis of the additional phosphate bond in ATP is about 7·7 kcal (32·2 kJ) per mole; thus, 15·4 kcal (64·5 kJ) are gained from metabolism of 1 mole of hexose.

Besides the increase in useful energy, the fermentation of carbohydrates is exothermic, a certain amount of energy being lost to the environment as heat. In a fermenter, this release of heat will produce a significant increase in temperature depending on the ratio of surface area (from which heat is dissipated) to volume (in which heat is generated). Thus the fermentation of a 22 % solution of sugar may give approximately an 11°C rise in temperature in a medium sized fermenter, and up to 22°C in a very large fermenter. In all except the smallest scale fermentations, some way of dispersing this heat has to be found, to ensure that the temperature does not reach 37°C at which, in general, fermentation ceases.

The theoretical yield of ethanol is 51·1 % by weight of the glucose metabolized. However, during the early, aerobic stage of yeast growth, 1–2 % of the sugar is completely oxidized to CO_2 and water. For this and other reasons, the practical yield of ethanol is never more than 48 % and can be much less.

Glycerol formation

Acetaldehyde has to be produced as a hydrogen acceptor if the overall reaction is to yield ethanol. Early in the fermentation, acetaldehyde is produced in small amounts only; dihydroxyacetone phosphate meanwhile acts as the hydrogen acceptor, becoming reduced to glycerol, which remains in small quantities in the final product. Glycerol formation (Figure 7.2) is important in certain white wines, on which it confers additional sweetness and smoothness.

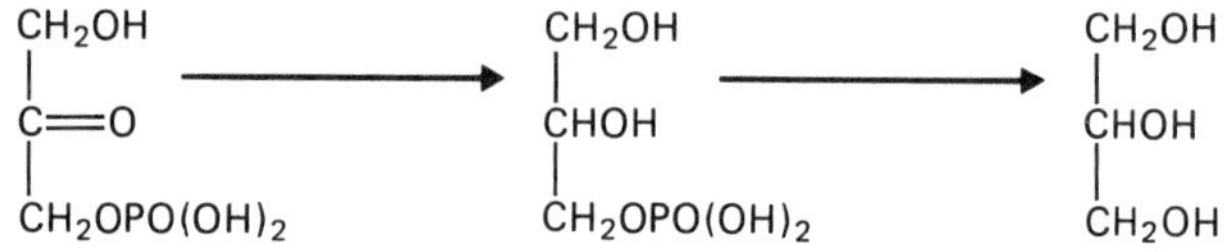

$$
\begin{array}{ccccc}
\text{CH}_2\text{OH} & & \text{CH}_2\text{OH} & & \text{CH}_2\text{OH} \\
| & \longrightarrow & | & \longrightarrow & | \\
\text{C}=\!\!=\!\!\text{O} & & \text{CHOH} & & \text{CHOH} \\
| & & | & & | \\
\text{CH}_2\text{OPO(OH)}_2 & & \text{CH}_2\text{OPO(OH)}_2 & & \text{CH}_2\text{OH}
\end{array}
$$

dihydroxyacetone   glycerol-1-phosphate   glycerol
phosphate

Figure 7.2

Other side-products of alcoholic fermentations

Small quantities of higher alcohols may be formed during deamination of amino- acids, particularly leucine and isoleucine, see Figure 7.3.

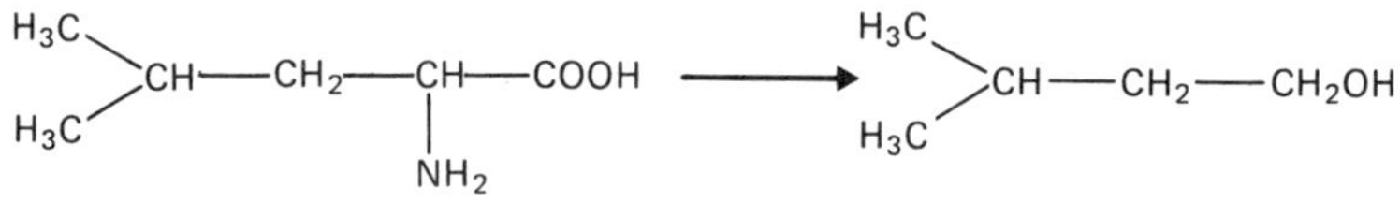

$$
\begin{array}{c}
\text{H}_3\text{C} \\
\quad\diagdown \\
\qquad\text{CH}-\text{CH}_2-\text{CH}-\text{COOH} \\
\quad\diagup \qquad\qquad\quad | \\
\text{H}_3\text{C} \qquad\qquad\quad \text{NH}_2
\end{array}
\longrightarrow
\begin{array}{c}
\text{H}_3\text{C} \\
\quad\diagdown \\
\qquad\text{CH}-\text{CH}_2-\text{CH}_2\text{OH} \\
\quad\diagup \\
\text{H}_3\text{C}
\end{array}
$$

leucine   *iso*-amyl alcohol

Figure 7.3

These higher alcohol fractions, together with the esters which they may form, are collectively termed 'fusel oils'. They are generally more toxic than ethanol, and amyl alcohol is the most toxic of all. Fusel oil formation is characteristic of grain fermentations rather than of fruit fermentations, on account of the higher protein content of grains.

Methanol formation, on the other hand, is more characteristic of wines, in which it derives from the pectins present in the fruit. For this reason, red wine (which is usually fermented in the presence of the pectin-containing grape skins) tends to contain more methanol than white wine.

Other side products, accounting for some 0.5–1.25% of the total sugar originally present, include lactic, succinic and acetic acids, 2-3-butanediol, acetaldehyde, acetylmethylcarbinol and diacetyl.

Process details for the production of alcoholic drinks are full of interest, but tend to contain much information which, though important, is non-microbiological. For this reason, the only two processes which will be considered in detail are those for the production of wine and barley-based beer. Production of spirits (whiskey, brandy, gin and rum) by distillation of *Saccharomyces*-fermented substrates is of insufficient microbiological interest for inclusion; however, it should be noted that in the whiskey fermentation, lactic acid bacteria play an important part, producing lactic, capric and caprylic acids which later produce flavour conferring esters.

7.2 Wine

Strictly defined, wine is a drink produced exclusively from the fermented juice of fresh wine grapes; various wine producing countries have legal definitions expressed in some such terms which, nonetheless, allow of certain additions during production. Wine is characteristically a European drink, and France and Italy still produce about half of the total world output of some 2×10^9 gallons (9×10^9 l) per annum.

The world's wine growing regions lie within belts of the North and South hemispheres where the mean annual temperature is 10–20°C. The ideal climate provides a long growing season with sufficient summer warmth for sugar production in the ripening fruit but not so much that the natural acidity is impaired; wines produced in warmer climates (for example, southern Spain, Cyprus and South California) tend to lack acidity.

Vitis vinifera, in its many hundreds of cultivars, is the principal wine grape, but *V. labrusca*, *V. rotundifolia* and others may be grown where a high summer humidity would make *V. vinifera* susceptible to spread of disease. Important *V. vinifera* cultivars include Sauvignon, the principal grape of red Bordeaux wine, Pinot Noir, the principal red Burgundy grape (Pinot Blanc is used for white Burgundy), Riesling and Sylvaner, widely used in German wines, Barbera and Freisa, used in the wines of Northern Italy and Palomino, the main sherry grape. Soil quality has a subtle influence on the final product: humus-rich soils do not produce the best wines, but wine grapes grow successfully on gravelley, slatey or even chalky soils, each exerting an influence on the nature of the wine. Vineyards tend to be on hillsides and even the position on the hill affects quality.

During ripening, the sugar content (glucose and fructose) of the grapes increases, acid content, particularly that of malic acid, falls and, in the case of dark grapes, pigmentation of the skins increases. Periodically, samples of some 200 berries are weighed, crushed and sugar content and total acidity determined. For highest productivity, harvesting begins when weight × sugar content reaches a maximum: for higher quality, balance between sugar content and total acidity is more important than maximum yield of wine. Harvesting occupies several weeks, branches being collected as they ripen.

The gathered grapes are crushed, usually mechanically, in such a manner that the unpleasant-tasting seeds are not broken: the juice yield is about 140 gallons per ton (0·6 l per kg). Red wines are produced from dark grapes, traditionally by fermentation of the entire product of the crushing, juice, pulp, skins, seeds and even sometimes the strongly tannin containing stalks; the anthocyanin pigments which confer red colour occur in the skins and are extracted during fermentation by the alcohol produced. Alternatively, there may be an initial heat extraction of the pigment by holding the crushed grapes at 45–75°C while juice is circulated over the pulp, or an 8–10 days maceration of the entire grapes in CO_2 filled vessels, during which an initial fermentation and pigment extraction occurs. For cheap red wines, the aim is maximum extraction of skin pigments without too much extraction of tannins which, though essential to the character of red wines, demand an extended maturation period.

White wines can be produced from either red or white grapes, but the pressed juice alone is sent to the fermenter and no pigment extraction occurs; extraction of tannin from the skins and seeds is also minimal.

The juice supplied to the fermenters is called must, whether or not it contains skins, seeds and pulp. The fermenting room of a wine plant usually contains tanks of various sizes to accommodate the different types of wine being produced and is in use for a short period each year following the grape harvest. The total fermentation capacity needed FC, in gallons, is given by an empirical formula

$$FC = \frac{100 \times DT \times FS}{4 \times TO}$$

where DT = daily tonnage of grapes, FS = length of the fermentation season, in days, and TO = turnover, i.e. the number of days the material remains in the fermenter.

Some details of must composition are given in Table 7.2. Insufficient sugar gives a wine of low alcoholic strength, susceptible to bacterial attack, and can be corrected by adding sucrose or by incorporating passulated (partly sun-dried) grapes or even raisins, practices which are restricted in certain countries. Sufficient acid content is extremely important: the acids produce a low must pH which allows yeast growth but inhibits most bacteria, they prevent development of various off-flavours during fermentation, they confer an agreeable tartness to the wine and finally, during maturation, they contribute to the formation of pleasant smelling and tasting esters. A low acid content can be corrected by adding tartaric or other acids, or calcium sulphate or phosphate, which liberate free tartaric acid from bitartrate:

$$2KH(C_4H_4O_6) + CaSO_4 \rightarrow Ca(C_4H_4O_6) + H_2(C_4H_4O_6) + K_2SO_4$$

Such additions are likewise restricted in some countries, such as W. Germany, and calcium sulphate treatment can readily be detected by analysis of the wine for sulphate content.

Fresh must contains live *S. ellipsoideus* and other yeasts which inhabit the waxy bloom on the surface of grapes (up to 10^5 cells per berry). If the must ferments naturally, a sequence of yeast populations may develop, perhaps initially species of *Kloeckera*, *Hanseniaspora* and *Candida*, then *Saccharomyces* species, including *S. ellipsoideus*, and *Torulospora* and finally other *Saccharomyces*, each successive population being destroyed by the mounting concentration of ethanol. An initial fermentation by *Kloeckera apiculata* and *Torulopsis* species is said to be characteristic of some Bordeaux wines. Such a mixed fermentation may produce a beneficial complexity of flavour; equally, it may lead to production of H_2S, mercaptans and other undesirable substances.

In practice, nearly all musts are treated with SO_2 to control or abolish the natural fermentation, by adding liquid SO_2 stored under pressure, SO_2 solutions or $KHSO_3$ (potassium bisulphite) or $K_2S_2O_5$ (potassium meta-bisulphite): this leads to the development of mixtures of dissolved SO_2, sulphurous acid H_2SO_3 and bisulphite ions in the must, all of which have antimicrobial action. 100 ppm SO_2 addition produces about a thousand-fold reduction in the number of viable cells in the must, killing most wild yeasts and bacteria but allowing *S. ellipsoideus* to survive, presumably because strains indigenous to wine growing areas have acquired resistance through repeated exposure. The restraining effect of the SO_2 addition quickly disappears through oxidation of H_2SO_3 and bisulphite to sulphate and through formation of bisulphite addition compounds with sugars and, after a slight delay, fermentation begins.

Alternatively, in large scale commercial wine production the must may be virtually sterilized by a heavier SO_2 addition, aerated to remove some SO_2, and a 1–3 % starter inoculum added. The starter consists of a pure wine yeast culture, selected for SO_2 resistance, ability to sediment after fermentation and appropriate flavour contribution; for example, a selected Burgundy, Montrachet or Champagne strain might be chosen. The starter is worked up in about ten-fold increasing volumes, beginning with pasteurized must and using sulphited must for the larger volumes. Choice of yeast inoculum exerts a subtle but definite influence on the final character of the wine, though must composition is the major factor.

Control of fermentation temperature is very important. The optimum temperature for rapid fermentation by most wine yeasts is 25–29°C, but there are cold-tolerant strains which ferment at 10°C; a rise of temperature to 35–37°C stops the fermentation, usually permanently, in which cases a fresh starter would have to be added. Red wine is commonly fermented at a higher temperature than white, up to about 27°C, which promotes extraction of pigment and tannin; there is some loss of ethanol by evaporation and entrainment in the CO_2 bubbles, and other volatiles are dissipated. White wines are generally fermented at 15–23°C, sometimes lower, which conserves natural fruit volatiles. Large wine fermenters are designed to present a large

Table 7.2 Summary of major
components of grape musts

Substance	*%*
Glucose	8–13*
Fructose	7–12*
Sucrose	(trace)
Tartaric acid	0·2–1·0
Malic acid	0·1–0·8*
Citric acid	0·01–0·05
Oxalic, gluconic, glucuronic, phosphoric acids	(traces)
Tannins/tannic acid	0·05–0·4**
Catechols	(traces)
Amino acids and other organic nitrogen	0·03–0·17
Ash	0·2–0·6
pH	3·1–3·9

* depends on ripeness
** more if the must contains skins, seeds or
stalks

surface area for heat dispersal and may have a cooling system, such as cooling coils within the fermenter or an arrangement for pumping screened must through a plate or cascade heat exchanger.

In red wine fermentation, a crust or 'pomace' of skins and pulp rises to the surface through CO_2 evolution. Extraction of pigment and tannin from the pomace is aided by pumping screened juice from the lower tank and spraying it over the crust; this also discourages the growth of obligately aerobic *Acetobacter* species in the crust. On a small scale, the pomace is manually submerged by repeated poling. If significant amounts of acetic acid form, indicating contamination by ethanol-oxidizing organisms such as *Acetobacter* or *Brettanomyces* species, remedial measures such as heavy SO_2 treatment are indicated, and a further inoculum may be needed.

Dryness or sweetness of the wine depends on a number of factors. Where the must has a high sugar content, fermentation slows and eventually stops short of total conversion and the wine remains sweet. Sweet wine can also be produced by arresting fermentation by a heavy addition of SO_2 (up to 1000 ppm) whilst there is still sugar present. Glycerol formation contributes to sweetness, and is encouraged by SO_2 addition early in fermentation, because the sulphite complexes with acetaldehyde, prolonging the dihydroxyacetone phase of fermentation.

A special approach to the production of sweet white wines is found in the Sauternes and Rhineland. Grapes infected with the mould *Botrytis cinerea* are reserved for production of high quality sweet wines. The fungus develops on the grape surface as a greyish coating ('pourriture noble' or 'Edelfäule' meaning 'noble rot'). Its mycelium penetrates the skin, abstracting some sugar and acids from the pulp for growth but promoting marked water loss. The grapes shrivel, sugar content rising to as much as 60%, and the resultant wine stops fermenting at a high residual sugar content. Attempts have been made to counterfeit 'pourriture noble' by artificially inoculating gathered grapes with *Botrytis* cultures.

Proline, the most plentiful must amino acid, is not metabolized by yeast and remains in the wine at up to 0·05%. Deamination of leucine and *iso*-leucine leads to formation of higher alcohols, including *iso*-amyl alcohol, and small quantities of succinic acid are formed during deamination of glutamic acid.

Cellar treatment

At the end of active fermentation, the wine is run into storage vats or tanks. During storage, various deposits form. Yeast cells sediment. As the wine cools, potassium hydrogen tartrate, representing about half of the original tartaric acid content of the must, forms heavy crystalline deposits on the walls and bottom of the container, and acidity is reduced. At a later stage, white wines are often held at 0°C for several weeks to precipitate tartrates which might later cause clouding of chilled wine. Also, during storage, positively charged pulp proteins slowly precipitate with negatively charged tannins.

Because the organic deposits slowly release unwanted flavours, the wine is racked into fresh containers after 1–2 months and on several subsequent occasions. To discourage contaminants and prevent flavour oxidation, the SO_2 content is generally brought up to 50–150 ppm before racking and white wines, which are particularly susceptible to flavour oxidation, may be treated with ascorbic acid or racked into CO_2 filled casks. The lees remaining after racking are either bulked, racked and the supernatant added to the wine, or used for distillation.

Red wines usually clarify rapidly because of their high tannin content. White wines clarify more slowly and may require the addition of a clarifying agent ('finings'), usually after the second racking. The usual fining agent is bentonite, a natural clay negatively charged in suspension, which is either added in amounts indicated by a laboratory test clarification (usually 0·1–0·2%) to give a slight precipitate which is filtered off together with the pulp debris, or in much larger quantities, and allowed to settle. Diluted egg albumin or blood serum can also be used.

Malo-lactic fermentation

After 10–300 days storage, depending on acidity, many wines begin to undergo a spontaneous bacterial fermentation caused mainly it would seem by growth of *Leuconostoc* strains, termed *L. oenos*, which can multiply in up to 10–13% ethanol and at pH values as low as 2·9–3·0. *L. oenos*, in common with many other lactic acid bacteria, produces the so-called malo-lactic enzyme which converts the dibasic malic acid directly to the monobasic lactic acid (Figure 7.4).

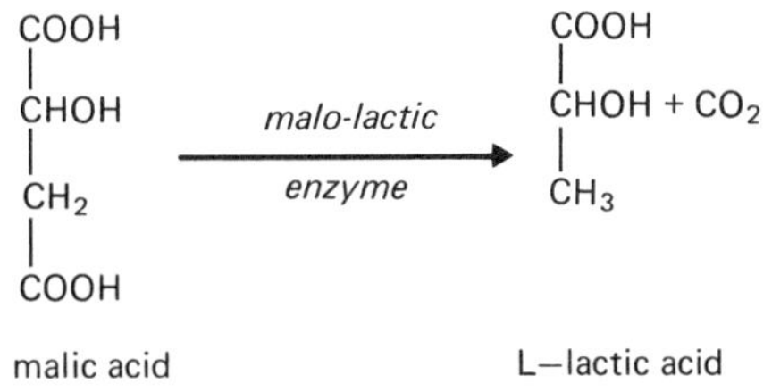

$$
\begin{array}{ccc}
\begin{array}{c}
\text{COOH} \\ | \\ \text{CHOH} \\ | \\ \text{CH}_2 \\ | \\ \text{COOH}
\end{array}
& \xrightarrow[\textit{enzyme}]{\textit{malo-lactic}} &
\begin{array}{c}
\text{COOH} \\ | \\ \text{CHOH} + \text{CO}_2 \\ | \\ \text{CH}_3
\end{array} \\[1em]
\text{malic acid} & & \text{L--lactic acid}
\end{array}
$$

Figure 7.4

The wine loses acidity, mellows in taste and, if bottled at this stage, exhibits slight effervescence (an example is the well known Mateus Rosé of Portugal). Less than 0·1% lactic acid in a wine indicates absence of malo-lactic fermentation, more than 0·2% indicates a high degree of fermentation.

Malo-lactic fermentation is considered desirable in many of the wines of Burgundy, the Rhineland and of Piedmont (where the acid Barbera and Freisa grape cultivars predominate) and is encouraged by refraining from SO_2 treatment and by storage at temperatures as high as 18°C. However it frequently also occurs in wines of already low acidity, especially if the ethanol content is also low, and here it is undesirable. It can be prevented by SO_2 treatment or, apparently, by adding 0·08–0·15% of fumaric acid to the young wine; fumaric acid has a bactericidal effect, but cannot be added to the must since it is metabolized by yeast during fermentation.

Ageing

Inexpensive wines are filtered and bottled as soon as deposition is judged complete or near-complete. Higher quality wines are aged for 1–4 years, normally in 225 litre wood casks, before bottling. The traces of oxygen which enter the wine through the cask walls and from the small air space in the cask cause desirable condensation, polymerization and precipitation of the tannins of red wines, lessening harshness; too much causes flavour oxidation and acetaldehyde formation and may allow growth of contaminants.

Partial esterification of ethanol and other alcohols with organic acids occurs, with fairly rapid formation of volatile ethyl acetate and lactate (the former has a crude, distinctive aroma and its flavour is discernible at a taste threshold of about 33 mg/l) and slower production of ethyl tartrate and succinate, which contribute to flavour. Esterification monitoring gives a convenient measure of the progress of ageing but gas-chromatography studies show that there is a concomitant production of other trace volatiles in very great variety, which markedly influence the final character of the wine.

Sparkling wines

Sparkling wines are variously made. The slight petillance produced by the malo-lactic fermentation has already been mentioned. Champagne, produced in the Rheims district, is made from a still, dry, acid white wine to which sucrose, a little tannin and a special strain of *S. ellipsoideus* of high CO_2 resistance and able to form a granular sediment are added at the time of bottling. After a secondary fermentation in the bottle at 15·5–18°C, the resulting yeast deposit is gradually eased to the cap end of the bottle by a progressive daily up-ending and shaking (remuage). The cap is then removed, the yeast plug (usually frozen by immersing the bottle neck in liquid CO_2) allowed to eject (dégorgement), a small sweetening dosage of sucrose, sometimes with brandy to inhibit further fermentation, added, and the bottles finally corked. Production of CO_2 in the bottle leads to the formation of an unstable ethyl pyrocarbonate which, on pouring the wine, gradually decomposes giving the long-sustained sparkle of true Champagne. Cheaper sparkling wines are made by secondary fermentation in closed tanks and the cheapest by injecting CO_2 under pressure into still wines: in this case, little ethyl pyrocarbonate is formed and the wine has only a brief sparkle in the glass.

Fortified wines

These are wines to which additional ethanol (brandy) is added after fermentation, usually raising the concentration to about 20%.

The most interesting microbiologically is sherry. Made from sun-dried, sweetish, low-acid Palomera grapes, sherry musts are generally plastered with $CaSO_4$ to increase acidity and little or no SO_2 is used. The new wine is stored in casks with a large air space (18–20%) to encourage secondary growth of highly aerobic, pellicle-forming yeasts, sometimes termed *S. beticus*, but probably *S. oviformis*, which utilize ethanol and glycerol, producing aldehydes, 2,3-butanediol and other compounds which confer the characteristic sherry flavour (sherry is one of the few wines in which a high degree of oxidation is actually desirable). The resulting yeast film ('flor') develops slowly, sometimes not at all: only those batches which develop flor are blended to form dry fino sherry, the rest being used for the sweeter sherries. 15–16% ethanol in the casked wine is optimal for flor development. Sherry blending is practised by the solera system: the butts are stored in tiers, each tier representing a year; annually, about $\frac{1}{4}$ of the contents of each butt is transferred to the next lowest tier and replaced by newer wine, injected below the surface so as not to disturb the flor film; sherry from the bottom tier is blended, fortified with brandy and bottled. Spanish oenologists have recently been studying the rapid induction of sherry flavour by bubbling sterile air through young wine in the presence of flor yeasts.

Port is a strong red wine in which primary fermentation is arrested while ample sugar is still present by adding alcohol or brandy; tannin precipitation may continue in the bottle. In Madeira and similar wines, a special flavour is obtained by heat treatment of the fermented wine; extra alcohol is then added. Vermouths are red or white wines flavoured with wormwood (German: vermuth) and other herbs, and fortified with alcohol or brandy.

7.3   Ale, beer and lager

These are drinks made by mashing barley malt and sometimes other cereals with water to produce a sugary wort, which is then fermented. The terms ale and beer have altered in usage. Anglo-Saxon ale was a fermented malt beverage, drunk fresh: when hopping was introduced in Britain in the 16th century, the resulting bitter drink with keeping qualities derived from hop resins was called 'beer'. Now, virtually all malt liquors in the Western world contain some hop extract, and the brewer uses the term 'beer' for his final product, as opposed to the unfermented 'wort': ale is a beer produced, generally, by top fermentation with *Saccharomyces cerevisiae* whilst lager is a beer made by bottom fermentation with *S. carlsbergensis*.

Malt

Barley malt is produced from low-N strains of barley. The dried grain is steeped and later spread on the malthouse floor. Warmth induces sprouting in 4–6 days, and amylolytic and proteolytic enzymes are formed. The sprouted grain is kilned by slow heating to about 80°C, much of the required enzyme activity surviving. The dried rootlets are knocked off and the remaining grain coarsely ground.

Water quality

Water ('liquor') quality for brewing is determined largely by the dissolved salts. Sulphates ($CaSO_4$, $MgSO_4$) are conducive to a dry, clean flavour in the final beer. $CaSO_4$ reacts with malt phosphates during mashing, increasing the acidity, and its presence is desirable when pale, non-acidic malts are used; it aids both protein coagulation during wort boiling and yeast flocculation at the end of fermentation. Burton-on-Trent water has a high total salts content (about 1200 ppm), with calcium sulphate predominating, and ale liquors are frequently 'burtonized' by adding $CaSO_4$ or a mixture of $CaCl_2$ and Na_2SO_4, sometimes with $MgSO_4$, to raise the permanent hardness to the equivalent of about 300 ppm $CaCO_3$.

Carbonate containing liquors are suitable for dark beers, balancing the higher acidity of darker malts. Carbonate increases extraction of hop bitters during boiling and dark beers are therefore lightly hopped. High carbonate is undesirable for more generously hopped light beers, producing a harsh bitterness. Chloride (mainly $NaCl$) is an important liquor constituent for dark, sweet ales, Dublin stouts excepted.

For lager, a low salts concentration is usually desirable, with a low temporary hardness (less than 20 ppm $CaCO_3$); Pilsen water contains only about 30 ppm total dissolved salts.

In the traditional brewery, which operates on a gravity principle, liquor treatment and heating takes place on a high level floor; during subsequent stages, the product passes downwards, floor by floor.

Mashing

The malt is mixed with coarsely ground, starchy cereals (rice, maize, wheat) to produce a grist which is cheaper and of lower N content. Mashing consists of adding hot water to the grist, holding for a period to allow enzymic conversions to occur, and draining off the resultant sweet wort.

Table 7.3 Properties of malt enzymes

Enzyme	Optimum temperature	Optimum pH	Principal mashing effect
α-amylase	60–70°C	5·7–6·3	dextrinization
β-amylase	57–65°C	5·0–5·5	saccharification
Proteinase	38–50°C	4·0–5·0	proteolysis

The principal enzymes involved are listed in Table 7.3 together with their optimum temperatures. The amylases together break down the straight glucose chain amyloses and the branched glucose chain amylopectins of starch to a mixture of maltose, maltotriose, maltotetraose and higher polymers of glucose (dextrins). Brewing yeasts readily hydrolyze and ferment the sugars up to maltotriose. A higher mashing temperature favours production of higher oligosaccharides and dextrins by α-amylase action; being non-fermentable, these remain in the final beer, giving sweetness, richness and foam retention. A lower mashing temperature favours saccharification by β-amylase, leading to a more fully fermented, carbohydrate-attenuated beer. Malt proteinases have temperature optima considerably lower than the amylases. Adequate proteolysis during mashing is important to produce peptides and amino acids for yeast nutrition and to reduce risk of development of protein hazes in the final beer, particularly when chilled. Choice of mashing temperatures, and other conditions, thus gives the brewer considerable control over the final product.

Infusion mashing

In this method, employed for British ales, the grist is mixed in the mash tun with hot liquor giving a mash temperature of about 65°C and stood for about 2 hours. At this temperature, combined α- and β-amylase action results in about 75% saccharification of the starch to fermentable sugars, the remainder forming slowly or non-fermentable oligosaccharides, and dextrins. Little proteolysis occurs. In one of many variants, there is an initial 'protein rest' of about 1 hour at a mashing temperature of 40–50°C, during which proteolysis occurs; then, boiled starchy adjuncts are mixed in to give a saccharification temperature of 65–70°C.

Decoction mashing

This method is more characteristic of bottom-fermented, lager-type beers. Again, there are many variants. Usually, hot liquor is mixed with the grist to give an initial mash temperature of about 40°C, and there is a protein rest of about 1 hour. The mash temperature is then stagewise raised by removing and boiling a proportion of the mash and returning it to the mash tun. For example, there might be a second protein rest at 50°C, a saccharification rest at 60–65°C and a dextrinization rest at 70°C.

At the end of all mashing processes, enzymic action is terminated by briefly raising the temperature to 75–80°C, which destroys the enzymes.

Initial mash pH is usually about 5·8, falling to about 5·4. Adjustment of mashing pH can give some measure of control, pH 5·5 favouring saccharification, 5·0 or less favouring proteolysis; lactic, phosphoric and sulphuric acids have been used as mash acidifiers.

After mashing, the sugary 'sweet wort' is filtered off through the grist residue which has settled onto the false bottom of the mash tun. The spent grist is leached ('sparged') with hot liquor to extract its remaining solubles.

Boiling

Sweet wort and spargings are mixed in the hop boiler (hop kettle; hop copper) and boiled for $1\frac{1}{2}$ to $2\frac{1}{2}$ hours during which hops are added, usually in several portions.

Hops are the dried female flower-cones of the climbing perennial, *Humulus lupulus*. The main ale hops in the U.K. are Goldings and Fuggles: the principal Continental lager hops are Saaz and Hallertauer and, in Germany, care is taken to ensure that no fertilization of the female flowers occurs so that the hops remain seedless.

Hop bitter acids are extracted during boiling and are important flavour constituents of beer; humulone (α-bitter acid) is more bitter than lupulone (β-bitter acid). During wort boiling, both α- and β-bitter acids convert to iso α- and β-bitter acids, which have stronger bitter flavours. Boiling also extracts negatively-charged hop tannins which coagulate with positively-charged wort proteins. Essential oils contributing to beer aroma (humulene, myrcene, linalool, geraniol) are also extracted, but largely boiled off; they may be restored by condensation, a proportion of the condensate being re-added to the boiled wort.

Early in wort boiling, a 'protein break' occurs: greyish curds of protein coagulated by heat and by hop tannins appear in the wort. This protein removal is important in prevention of haze formation at later stages.

The boiled wort is strained through the 'hop back', a crude filter at the bottom of the copper which utilizes the spent hops as a filter medium. The strained wort passes by gravity through a cooler and thence, at 12–20°C, to the fermenting vessels.

Batch fermentation

The fermentation vessels in batch brewing may be either skimmed, open tanks or closed cylindro-conical vessels. The latter form, the Nathan fermenter (Figure 7.5), has a system of cooling jackets which induces convection mixing of the contents, speeding fermentation, favouring eventual sedimenting of yeast and the protein-tannin complexes, and reducing turn-round time. Since the Nathan fermenter is closed, skimming is impossible and a sedimenting yeast must be used: the conical bottom of the fermenter favours maximum sediment packing. The Nathan fermenter also requires a yeast strain producing little foam during fermentation.

Yeast strains

Ale is brewed with top-fermenting strains of *S. cerevisiae*; a large proportion of the yeast crop is carried to the surface by CO_2 bubbles and the creamy foam is

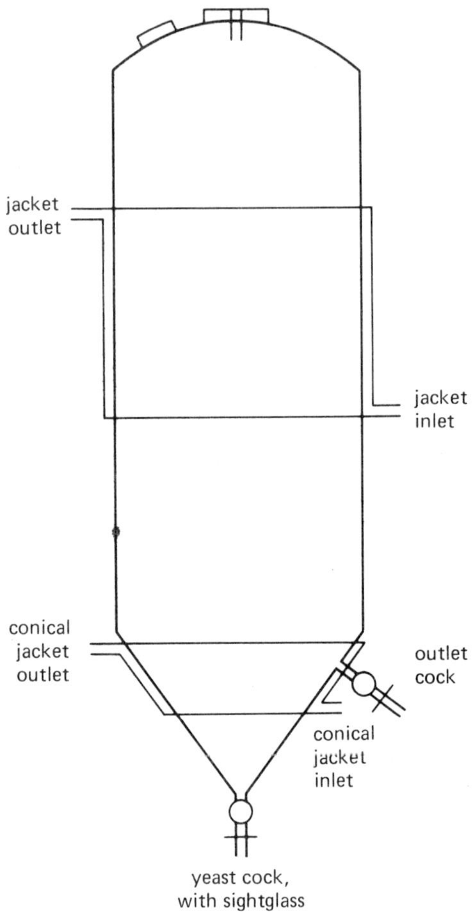

Figure 7.5   Cylindro-conical (Nathan) fermenter

skimmed off; at the end of fermentation, the remaining yeast sediments. The middle skimmings may be reserved to inoculate subsequent brews. Ale yeasts operate best at about 20°C and the fermentation time is about 5 days (3 days if Nathan fermenters are used).

For lagers, bottom-fermenting yeasts termed *S. carlsbergensis* are used. These produce less foam and the bulk of the yeast crop sinks early to the bottom of the fermenter; rousing of the sediment may be needed. Fermentation lasts 7–14 days at 5–15°C, and the middle layer of yeast sediment may be collected for re-use. In the UK Nathan fermenters are often used for lager fermentation.

Top and bottom fermenting yeasts tend to differ in their action on the trisaccharide raffinose (galactose + glucose + fructose). Top fermenters (which are usually spherical-celled) generally produce invertase, hydrolyzing the glucose-fructose part of the molecule, but are unable to hydrolyze the remaining melibiose (glucose-galactose); thus, only $\frac{1}{3}$ of the raffinose is fermented. Bottom fermenters generally produce both invertase and melibiase and thus ferment raffinose completely. The distinction is not very exact: for example, some ale yeasts are bottom fermenters.

Until recently, brewers re-used their yeasts indefinitely, reserving a portion of each yeast crop for pitching the next brew. With time, these yeasts tended to become highly mixed, containing both variants and, sometimes, wild yeast populations; they operated because of rigid adherence to precise brewing conditions. Pure culture brewing was first used for lager fermentation, but is now increasingly used for ale production. A selected, pure-culture yeast is carefully maintained in the laboratory and grown up in small batches or in a 'yeast machine' (a small continuous fermenter) for pitching purposes. Selection of a suitable strain may require screening of hundreds of cultures.

Evaluation of brewing yeast strains

Evaluation is usually by means of scaled-down trial fermentations of the hopped wort which it is proposed to use, in 2–3 litre tubular vessels of glass or steel, sometimes with a vacuum jacket. This can give information on yeast growth rate, amount of yeast head and deposit, degree and rate of sugar attenuation, total yeast crop, rate of yeast sedimentation at the end of fermentation and loss of isohumulone by yeast adsorption; also, preliminary quality tests may be carried out on the final beer. A few likely strains are then examined on a slightly larger laboratory scale, perhaps with successive fermentations to test yeast stability. The best candidate is then tested in a pilot fermentation of about 1500 litres, which seems to be the minimum quantity for successful flavour evaluation.

Various intrinsic characters of yeast strains can also be tested. Fermentation velocity is measured by the amount of CO_2 evolved per hour per gram of moist yeast, from a defined substrate. Fermentation efficiency is the ratio of fermentation velocity to yeast N content, measured by Kjeldahl digestion. Flocculation can be measured in calcium containing buffers, in fermented wort or in defined artificial beer: it is a highly important intrinsic yeast character, for the yeast must flocculate extensively in order that the beer should clear, but only at the end of fermentation. Other intrinsic characters include ability to ferment maltotriose, adsorption of hop bitters, fusel oil production, yeast oxygen requirement, attenuation limit and strain stability including rate of mutation to respiratory deficiency.

Brewing yeast genetics

A little is known about genetic control of some of the more important brewing properties. Ability to ferment maltose is associated with at least four dominant, polymeric genes, MAL-1, -2, -3 and -4, each enabling a characteristic level of the necessary α-glucosidase to be produced; strains possessing more than one dominant MAL allele form the enzyme in amounts near the expected cumulative value; only the homozygous recessive form fails to produce α-glucosidase. The actual rate of fermentation appears to be associated with ploidy, 3n and 4n forms fermenting faster than those of higher or lower ploidy. Flocculence (F) is dominant to non-flocculence (f) and, especially in top yeasts, F tends to mutate to f.

Many yeasts, brewing strains included, spontaneously mutate at a very low rate to a small-colony, small-cell form termed a 'petite' or respiratory-deficient mutant (RDM). Petites lack respiratory enzymes, notably cytochrome c oxidase and cytochromes a and b and thus form colourless colonies on triphenyl tetrazolium chloride agar (the parent forms producing red colonies). Laboratory culture in the presence of proflavine or acriflavine increases the rate of mutation to petite. In worts prepared from mashes treated with low levels of formaldehyde to reduce haze formation in the final beer, petite formation is also increased and, because of the slower sedimentation of the smaller cells, clearing of the beer may be delayed; moreover, excessive diacetyl may be produced, giving flavour spoilage.

Pure culture beer fermentations, both batch and continuous, require ideally a single yeast strain in which all desirable brewing properties are exactly combined. Hybridization might seem the obvious solution but in practice many brewing yeasts, particularly *S. carlsbergensis*, sporulate poorly if at all, or exhibit polyploidy, and the necessary haploids cannot be obtained. However, using mass-mating procedures, hybrids suitable for continuous fermentation have been produced, and also top/bottom yeast hybrids with improved flavour properties. In another instance, it proved possible to cross *S. carlsbergensis* with a dextrin-fermenting ('super-attenuating') yeast, *S. diastaticus*, giving a hybrid which produced highly attenuated but palatable beer, suitable for diabetics. Major yeast hybridization developments are still awaited by the brewing industry. Meanwhile, the brewing technologist is obliged to rely largely on manipulation of fermentation conditions, coupled with the high degree of phenotypic plasticity exhibited by the brewing yeasts.

Changes during fermentation

During fermentation, ethanol is formed together with other alcohols and volatiles termed 'fusel oil'; sugar content and specific gravity diminishes. The nitrogen content is reduced. Acetic, pyruvic, succinic, α-ketoglutaric, pyroglutamic, hydroxyglutaric, malic and citric acids are formed (total 350–750 mg/l) and the pH falls. A proportion of the iso-α-bitter acids is adsorbed onto the yeast cell walls. Some specimen values for a particular brewery are given in Table 7.4 but it should be understood that these are likely to vary widely. Near the end of fermentation, a spontaneous flocculation of yeast cells (the 'yeast break') occurs due to the formation of calcium or carboxylic bridges between adjacent cells, and the floccules sink leaving the beer almost clear.

Post-fermentation processing

The 'green' beer is run into storage vessels where it is conditioned to remove its harsh flavour. Ale is conditioned at 0–10°C for a period which may be very brief or up to several weeks. Lager is stored at 0–3°C for 3 weeks up to several months, during which further slow fermentation and protein precipitation occurs. One of the early difficulties with pure ale yeast cultures was their lack of conditioning ability.

Fining, the addition of flocculating agents such as isinglas, is now little used. Instead, the beer is passed through a continuous centrifuge to remove remaining suspended cells. It is then carbonated to give a CO_2 content of about 0·5%, filtered to give brilliance and filled into bottles, cans or pressure kegs. Pasteurization is common.

Continuous fermentation

Some beer is produced by continuous fermentation in the United States, New Zealand, Spain and other countries. In the UK about 6% of total beer production is by this method. The beer continuous fermenter is a chemostat, through which a slow flow of wort keeps the yeast at its phase of growth retardation.

The commonest form is the tower fermenter, a tall, cylindrical vessel about 40×3 feet (12×1 m), which is charged with a concentrated suspension of a slow-growing, highly flocculent yeast. Since the yeast population is almost static, a low-nitrogen wort is needed. This is pumped in at the bottom of the tower and the product emerges at the funnel-shaped top; typical residence times are 4 hours for ale, 8 hours for lager. Technical difficulties include the possible formation of large sedimented yeast plugs which are carried to the top by a pillow of carbon dioxide gas.

The advantages claimed for continuous beer fermentation are economy of plant space, rapid throughput, lower total labour costs, greater alcohol yield (because of retarded yeast growth) and uniformity of product. The main drawbacks are lack of flexibility making it difficult for the brewery to produce a range of different beers, high capital cost and the need for continuous wort production, beer finishing and packaging.

Table 7.4   Specimen analytical details for an ale and a
lager, before and after fermentation

	Ale		Lager	
Item	wort	beer	wort	beer
Sugars (g/100 ml)				
Fructose	0·512	0·165	0·063	0·005
Glucose	1·549	0·084	1·233	0·115
Sucrose	0·561	0·053	0·488	0·000
Maltose	3·000	0·143	7·295	0·231
Maltotriose	1·462	0·325	1·438	0·230
Nitrogen (mg/100 ml)				
Total soluble N	65·1	47·5	75·3	65·1
α-amino N	17·3	3·8	29·0	10·2
Fusel oils (ppm)				
n-propanol		8·1		5·1
Iso-butanol		15·2		3·9
Amyl alcohol		94·3		64·6
β-phenyl ethanol	(not done)	22·7	(not done)	5·2
Diacetyl		0·45		0·05
Acetaldehyde		4·2		10·6
Acetone		2·8		5·1
Ethyl acetate		19·5		20·6
Dimethyl sulphide				0·03
pH	5·52	4·10	5·17	4·26
S.G.	1·037	1·008	1·048	1·009

Beer quality control

Physical and chemical methods

Aroma producing volatiles (higher alcohols, esters, aldehydes, diacetyl, dimethyl sulphide, β-phenyl ethanol) are measured by gas liquid chromatography. Measurement of the specific gravity of the wort (original gravity, O.G.), the fermented beer (apparent gravity, A.G.) and, sometimes, of the beer after further fermentation to attenuation limit (A.L.) give an indirect measure of alcohol content (O.G. − A.G.) and of sweetness (A.G. − A.L.). Determination of refractive index and apparent gravity (A.G.), with reference to special tables, gives both sugar and alcohol content. Bitterness can be measured by solvent extraction of iso-α-acids and determination of UV absorption at 275 nm. Colour, pH, salt content, haze and head formation and retention may also be measured.

Microbiological methods

Routine cultural examination is carried out on the finished beer and, where pitching is by yeast from a previous brew, on the yeast inoculum. The commonest contaminants of skimmed open tanks are *Acetobacter* and *Acetomonas* species. Cylindro-conical fermentation vessels tend to favour contamination by *Lactobacillus* and *Pediococcus* species, but the fall in oxygen content in the closed vessels tends to make contamination by these organisms a dormant one. The source of the contaminants is usually the brewery air; *Lactobacillus* contamination in particular derives from the dust produced by the mills used to grind malt and barley.

Microbial beer contamination produces hazes, pellicles, acidification and undesirable flavours or consistency. *Acetobacter* species cause acidification and haze production. *Acetobacter xylinum* produces a polysaccharide slime leading to ropey consistency. Slow growing *Lactobacillus* species can produce acidity, haze and tangs in keg ales and lagers. *Pediococcus cerevisiae* is of greatest importance in lager storage, during which it can produce diacetyl, imparting a honey-like odour and taste. *Zymomonas* species, including *Z. anaerobica*, are characteristic stench-producers in dark, sweet beers. A variety of coliform bacteria can be isolated from beer, including the common *Obesumbacterium proteus*; this is probably a species of *Hafnia* which has become adapted to anaerobic wort conditions, under which it produces a parsnip like odour. Yeast contaminants include 'wild' yeasts of the genera *Pichia*, *Candida*, *Hanseniaspora* and *Rhodotorula* as well as *Saccharomyces* species other than *S. cerevisiae* or *S. carlsbergensis*, and non-brewing strains of the latter two. Wild yeasts can cause pellicle formation and off-flavours in stored beer.

Many semi-selective media are available for the detection of specific contaminants. A medium containing lysine as the sole nitrogen source is widely used for the detection of wild yeasts, which can grow on it. *Saccharomyces* species do not grow on lysine agar, but media containing critical concentrations of actidione (4 ppm), crystal violet or fuchsin sulphite have been described, which inhibit *S. cerevisiae* but allow other *Saccharomyces* to grow. Higher concentrations of actidione are frequently used in media for the detection of bacterial contaminants, to suppress yeast growth completely. Acetic acid bacteria can be selectively grown on beer/yeast extract/ethanol agar at pH 4·0 containing actidione. *Lactobacillus* and *Pediococcus* species are detected by growth on beer/malt extract agar with either thallous acetate or lactic acid, actidione and polymyxin added, and incubated in an atmosphere containing added CO_2.

Direct plating onto such media is adequate for examination of pitching yeast, and will detect infection down to about 30 organisms per ml of beer. For modern filtered beers, improved levels of detection are secured by membrane filtration of a large sample (< 1 litre), followed by incubation of the membrane on a suitable agar medium.

Conventional cultural techniques are too slow to be really suitable for the modern brewery with its rapid turnover. Rapid methods of contaminant detection which have been suggested but little used include ATP and NAD assay (sensitivity 10^3 cells/ml) pH and impedance changes and uptake of radioactively labelled substrate (10^4 cells/ml) and gravimetric and cell counting or staining methods (10^2 cells/ml). Immunofluorescence techniques are increasingly being used for the rapid detection of wild yeasts. Smears of pitching yeast or of filtered or centrifuged beer are stained with a fluorescein isothiocyanate conjugated rabbit antiserum against a selection of common wild yeasts, and examined by ultra-violet light microscopy: wild yeast cells exhibit a yellow-green fluorescence. Alternatively, microcolonies developing after about 6 h incubation of filter membranes on nutrient media can be fixed, stained with the fluorescent antiserum and the colonies counted under low-power incident ultra violet light microscopy.

Bibliography

AMERINE, M. A., (1964), 'Wine', *Scientific American*, Reprint 190.

AMERINE, M. A., BERG, H. W. and CRUESS, W. V., (1967), *The Technology of Wine Making*, 2nd ed., Avi Publishing Co. Inc., Westport, Conn.

AMERINE, M. A. and JOSLYN, M. A., (1970), *Table Wines: the Technology of their Production*, University California Press.

AULT, R. G., (1965), 'Spoilage bacteria in brewing', *J. Inst. Brewing*, **71**, 376–391.

BURNETT, J. H., (1975), *Mycogenetics*, John Wiley and Sons, Inc.

CARR, J. G., (1968), *Biological Principles in Fermentation*, Heinemann Educational Books, Ltd., London.

CARR, J. G., CUTTING, C. V. and WHITING, G. C., (ed.), (1975), *Lactic Acid Bacteria in Beverages and Food*, Academic Press, London and New York.

COOK, A. H., (ed.), (1962), *Barley and Malt*, Academic Press, New York and London.

COOTE, N. and KIRSOP, B. H., (1977), 'The content of organic acids in beer and other fermented media', *J. Inst. Brewing*, **80**, 474–483.

DUNCAN, J., (1973), 'Quality control in the brewery', *Process Biochemistry*, Feb. 1973, 14–16.

DUPUY, P., (1974), 'La chimie du vin', *La Recherche*, **5**, 837–845.

EMEIS, C. C., (1971), 'A new hybrid yeast for the fermentation of wort dextrins', *Proc. Amer. Soc. Brewing Chemists*, 51.

FINDLAY, W. P. R., (ed.), (1971), *Modern Brewery Technology*, Macmillan, New York.

HESSELTINE, C. W., (1965), 'A millenium of fungi, food and fermentation', *Mycologia*, **57**, 149–197.

HOUGH, J. S., BIGGS, D. E. and STEVENS, R., (1971), *Malting and Brewing Science*, Chapman and Hall, London.

LICHINE, A., (1963), *Wines of France*, Alfred A. Knopf.

PEDERSON, C. S., (1971), *Microbiology of Food Fermentations*, Avi Publishing Co., Inc., Westport, Conn.

STEWART, G. G., (1974), 'Some thoughts on microbiological aspects of brewing and other industries utilizing yeast', *Advances in Applied Microbiology*, **17**, 233–264.

UNDERKOFLER, L. A. and HICKEY, R. J., (ed.), (1954), *Industrial Fermentations*, (2 vols.), Chemical Publishing Co., Inc., New York.

WALKEY, R. J. and KIRSOP, B. H., (1969), 'Performance of strains of *Saccharomyces cerevisiae* in batch fermentation', *J. Inst. Brewing*, **75**, 393–398.

Index